Nonlinear Parameter Optimization Using R Tools

Nonlinear Parameter Optimization Using R Tools

John C. Nash

Telfer School of Management
University of Ottawa

Library of Congress Cataloging-in-Publication Data

Nash, John C., 1947-
    Nonlinear parameter optimization using R tools / John C. Nash.
        pages cm
    Includes bibliographical references and index.
    ISBN 978-1-118-56928-3 (cloth)
    1. Mathematical optimization. 2. Nonlinear theories. 3. R (Computer program language) I. Title.
    QA402.5.N34 2014
    519.60285′5133−dc23

                                                        2013051141

A catalogue record for this book is available from the British Library.

ISBN: 9781118569283

Typeset in 10/12pt TimesLTStd by Laserwords Private Limited, Chennai, India

1   2014

This work is a part of and is dedicated to that effort by the many community-minded people who create, support, promote, and use free and open source software, and who generously share these ideas, without which R in particular would not exist.

Contents

Preface

This book is about tools for finding parameters of functions that describe phenomena or systems where these parameters are implicated in ways that do not allow their determination by solving sets of linear equations. Furthermore, it is about doing so with R.

R is a computer language and suite of libraries intended primarily to be used for statistical and mathematical computations. It has a strong but quite small base system that is rich in functions for scientific computations as well as a huge collection of add-in packages for particular extra problems and situations.

Among both the base and add-in tools are facilities for finding the "best" parameters of functions that describe systems or phenomena. Such tools are used to fit equations to data or improve the performance of devices or find the most efficient path from A to B, and so on. Some such problems have a structure where we only need solve one or two sets of linear equations. Others – the subject of this book – require iterative methods to approximate the solution we desire. Generally we refer to these as nonlinear parameters. A formal definition will be given later, but the tools we shall discuss can be applied anyway.

Sometimes the problems involve constraints as well as objective functions. Dealing with constraints is also a subject for this book. When there are many constraints, the problem is usually called *mathematical programming*, a field with many subspecialties depending on the nature of the objective and constraints. While some review of the R tools for mathematical programming is included, such problems have not been prominent in my work, so there is less depth of treatment here. This does not imply that the subject should be ignored, even though for me, and also to some extent for R, mathematical programming has been a lower priority than nonlinear function minimization with perhaps a few constraints. This reflects the origins of R in the primarily statistical community.

The topics in this book are most likely to be of interest to researchers and practitioners who have to solve parameter determination problems. In some cases, they will be developing specialized tools, such as packages for maximum likelihood estimation, so the tools discussed here are like the engines used by car designers to prepare their particular offering. In other cases, workers will want to use the tools directly. In either case, my aim is to provide advice and suggestions of what is likely to work well and what should generally be avoided.

The book should also interest the serious users of composite packages that use optimization internally. I have noticed that some such packages call the first or most obvious tool in sight. I believe the user really should be offered a choice or at least informed of what is used. Those who stand most to benefit from the book in this way are workers who have large computationally intensive problems or else awkward problems where standard methods may "fail" – or more likely simply terminate before finding a satisfactory solution.

Thus my goal in preparing this book is to provide an appreciation of the tools and where they are most useful. Most users of R are not specialists in computation, and the workings of the specialized tools are a black box. This can lead to mis application. A lumberjack's chain saw is a poor choice for a neurosurgeon's handiwork. But a system like R needs to have a range of choices for different occasions. Users also need to help in making appropriate choices.

I also hope to provide enough simplified examples to both help users and stimulate developers. While we have a lot of tools for nonlinear parameter determination in R, we do not have enough well-organized guidance on their use or ways to effectively retire methods that are not good enough.

As a developer of such tools, I have found myself confronted on a number of occasions by users telling me that a colleague said my program was very good, but "it didn't work." This has almost always been because the mode of use or choice of tool was inappropriate. Ideally, I aim to have the software identify such misapplication but that is a very difficult challenge that I have spent a good deal of my career addressing.

That said, I must note that many many items that appear in this book were developed or adjusted during its writing. The legacy of a very large and widely used open source system leaves much detritus that is not useful and many inconvenient loose ends. Some have been addressed as the writing progressed.

A word about the structure of this book. Readers who read cover to cover will find a certain amount of duplication. I have tried to make each chapter somewhat self-contained. This does not mean that I will not refer the reader to other chapters, but it does mean that sometimes there is repetition of ideas that are also treated elsewhere. This is done to make it easier to read chapters independently.

Any book, of course, owes a lot to people other than the author. I cannot hope to do justice to every person who has helped with ideas or support, but the following, in no particular order, are the names of some people who I feel have contributed to the writing of this book: Mary Nash, Ravi Varadhan, Stephen Nash, Hans Werner Borchers, Claudia Beleites, Nathan Lemoine, Paul Gilbert, Dirk Eddelbeuttel, John Fox, Ben Bolker, Doug Bates, Kate Mullen, Richard Alexander and Gabor Grothendieck.

John C. Nash

1

Optimization problem tasks and how they arise

In this introductory chapter we look at the classes of problems for which we will discuss solution tools. We also consider the interrelationships between different problem classes as well as among the solution methods. This is quite general. R is only incidental to this chapter except for some examples. Here we write our list of things to do.

1.1 The general optimization problem

The general constrained optimization problem can be stated as follows.

> Find $\mathbf{x} = argmin\ f(\mathbf{x})$
> such that
> $\mathbf{c}(\mathbf{x}) >= 0$

Note that $f()$ is a scalar function but $\mathbf{x}$ is a vector. There may or may not be **constraints** on the values of $\mathbf{x}$, and these are expressed formally in the vector of functions $\mathbf{c}$. While these functions are general, many problems have much simpler constraints, such as requirements that the values of $\mathbf{x}$ be no less than some lower bounds or no greater than some upper bounds as we shall discuss in the following text.

We have specified the problem as a minimization, but maximization problems can be transformed to minimizations by multiplying the objective function by -1.

Note also that we have asked for the set of arguments $\mathbf{x}$ that minimize the objective, which essentially implies the global minimum. However, many – if not most – of the numerical methods in optimization are able to find only local minima

Nonlinear Parameter Optimization Using R Tools, First Edition. John C. Nash.
© 2014 John Wiley & Sons, Ltd. Published 2014 by John Wiley & Sons, Ltd.
Companion Website: www.wiley.com/go/nonlinear_parameter

and quite a few problems are such that there may be many local minima and possibly even more than one global minimum. That is, the global minimum may occur at more than one set of parameters **x** and may occur on a line or surface.

1.2   Why the general problem is generally uninteresting

While there do exist methods for tackling the general optimization problem, almost all the "real" work of optimization in problems related to statistics and modeling tends to be done by more specialized methods that work on problems that are restricted in some ways by the nature of the objective or the constraints (or lack thereof). Indeed, for a number of particular problems, there are very specialized packages expressly designed to solve them. Unfortunately, the user often has to work quite hard to decide if his or her problem actually matches the design considerations of the specialized package. Seemingly small changes – for example, a condition that parameters must be positive – can render the specialized package useless. On the other hand, a very general tool may be quite tedious for the user to apply easily, because objective functions and constraints may require a very large amount of program code in some cases.

In the real world, the objective function $f()$ and the constraints **c** are not only functions of **x** but also depend on data; in fact, they may depend on vast arrays of data, particularly in statistical problems involving large systems.

To illustrate, consider the following examples, which, while "small," illustrate some of the issues we will encounter.

Cobb–Douglas example

The Cobb–Douglas production function (Nash and Walker-Smith, 1987, p. 375) predicts the quantity of production prodn of a commodity as a function of the inputs of kapital (it appears traditional to use a K for this variable) and labour used, namely,

$$\text{prodn} = b_1 * \text{kapital}^{b_2} * \text{labour}^{b_3} \qquad (1.1)$$

A traditional approach to this problem is to take logarithms to get

$$\log(\text{prodn}) = \log(b_1) + b_2 * \log(\text{kapital}) + b_3 * \log(\text{labour}) \qquad (1.2)$$

However, the two forms imply very different ways in which errors are assumed to exist between the model and real-world data. Let us assume (almost certainly dangerously) that data for kapital and labour are known precisely, but there may be errors in the data for prodn. Let us use the name Dprodn. In particular, if we use additive errors of the form

$$\text{errors} = \text{data} - \text{model} \qquad (1.3)$$

then we have

$$\log(\text{Dprodn}) = \log(b_1) + b_2 * \log(\text{kapital}) + b_3 * \log(\text{labour}) + \text{errorsL} \qquad (1.4)$$

where we have given these errors a particular name errorsL. This means that the errors are actually multiplicative in the real scale of the data.

$$\text{Dprodn} = b_1 * \text{kapital}^{b_2} * \text{labour}^{b_3} * \exp(\text{errorsL}) \qquad (1.5)$$

If we estimate the model using the log form, we can sometimes get quite different estimates of the parameters than using the direct form. The "errors" have different weights in the different scales, and this alters the estimates. If we really believe that the errors are distributed around the direct model with constant variance, then we should not be using the log form, because it implies that the relative errors are distributed with constant variance.

Hobbs' weed infestation example

This problem is also a nonlinear least squares. As we shall see later, it demonstrates a number of computational issues. The problem came across my desk sometime in 1974 when I was working on the development of a program to solve nonlinear least squares estimation problems. I had written several variants of Gauss–Newton methods in BASIC for a Data General NOVA system. This early minicomputer offered a very limited environment of a 10 character per second teletype with paper tape reader and punch that allowed access to a maximum 8K byte (actually 4K word) segment of the machine. Arithmetic was particularly horrible in that floating point used six hexadecimal digits in the mantissa with no guard digit.

The problem was supplied by Mr. Dave Hobbs of Agriculture Canada. As I was told, the observations (y) are weed densities per unit area over 12 growing periods. I was never given the actual units of the observations. Here are the data (Figure 1.1).

```
# draw the data
y <- c(5.308, 7.24, 9.638, 12.866, 17.069, 23.192, 31.443, 38.558, 50.156,
    62.948, 75.995, 91.972)
t <- 1:12
plot(t, y)
title(main = "Hobbs' weed infestation data", font.main = 4)
```

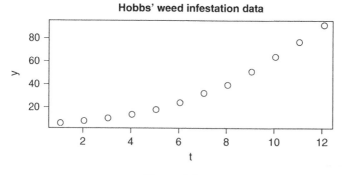

Figure 1.1

It was suggested that the appropriate model was a 3-parameter logistic, that is,

$$y_i = b_1/(1 + b_2 \exp(-b_3 t_i)) + \varepsilon_i \tag{1.6}$$

where $\varepsilon_i \sim N(0, \sigma^2)$, t_i is the growing period, and $i = 1, \ldots, 12$. We shall see later that there are other forms for the model that may give better computational properties.

1.3   (Non-)Linearity

What do we mean by "nonlinear?" The word clearly implies "not a straight line," and many researchers take this to apply to the model they are trying to estimate. However, for the process of estimation, which generally involves minimizing a loss function such as a sum of squared deviations or maximizing a likelihood function, the key issue is that of solving a set of equations to find the result.

When we minimize the sum of squares for a model that is linear in the parameters, such as the log form of the Cobb–Douglas function (1.2) above where $\log(b_1)$, b_2, and b_3 appear only to the first power, we can apply standard calculus to arrive at the *normal equations*. These are a set of linear equations. However, when we want to minimize the sum of squares from the original model (1.1), it is generally necessary to use an iterative method from some starting set of the parameters b_1, b_2, and b_3.

For the purposes of this book, "nonlinear" will refer to the process of finding a solution and implying that there is no method that finds a solution via a predetermined set of solutions of linear equations. That is, while we use a lot of linear algebra in finding solutions to the problems of interest in this book, we cannot, in advance, specify how many such subproblems are needed.

1.4   Objective function properties

There are some particular forms of the objective function that lead to specialized, but quite common, solution methods. This gives us one dimension or axis by which to categorize the optimization methods we shall consider later.

1.4.1   Sums of squares

If the objective function is a sum of squared terms, we can use a method for solving **nonlinear least squares** problems. Clearly, the estimation of the Cobb–Douglas production model above by minimizing the sum of squared residuals is a problem of this type.

We note that the Cobb–Douglas problem is linear in the parameters in the case of the log-form model. The linear least squares problem is so pervasive that it is worth noting how it may be solved because some approaches to nonlinear problems can be viewed as solving sequences of linear problems.

1.4.2   Minimax approximation

It is sometimes important to have an upper bound on the deviation of a model from "data." We, therefore, wish to find the set of parameters in a model that minimizes the maximum deviation, hence a *minimax problem*. In particular, consider that there may be relatively simple approximations to some specialized and awkward-to-compute special functions. This sort of *approximation problem* is less familiar to statistical workers than sums-of-squares problems. Moreover, the small residuals may render some traditional methods such as the R function `nls()` ill-suited to their solution.

1.4.3   Problems with multiple minima

Possibly as a result of the general mathematical education most of us receive, our view of functions is that they have a "nice" shape where there is just one minimum or maximum. Reality is much less kind. Functions often have multiple local extrema, and most of our methods – often based on traditional calculus ideas – are designed to find local minima or maxima efficiently. Knowing that there may be multiple local, and possibly even multiple global, solutions is important if we are to find the right solution or even an acceptable solution.

1.4.4   Objectives that can only be imprecisely computed

Sometimes an objective function is not precisely determined. On one occasion, I considered optimizing racing-car parameters such as fuel–air ratio, braking settings, and so on by "computing" the objective via measurement of the time taken to complete a circuit of the track. This is clearly stochastic. There are other such problems where integrals are estimated by Monte-Carlo methods. See, for example, http://en.wikipedia.org/wiki/Monte_Carlo_integration for a general overview.

1.5   Constraint types

Another categorization of optimization problems and their solution methods is via the constraints that are imposed on the parameters.

The most obvious case is **unconstrained** – no constraints. Truthfully, real problems always have constraints, but many can be most easily and efficiently solved without the need to explicitly include them.

The next simplest is **bounds** or **box constraints**. Here we impose lower and upper bounds on the parameters. If we have lower bounds in a vector **lo** and upper bounds in a vector **up**, then the parameters are in the n-dimensional box defined by the inequalities

$$\mathbf{lo} \leq \mathbf{x} \leq \mathbf{up} \tag{1.7}$$

Linear constraints take the form

$$\sum z_i * x_i \le q_i \qquad (1.8)$$

or

$$\mathbf{z}' * \mathbf{x} \le \mathbf{q} \qquad (1.9)$$

If all constraints are linear (which includes bounds), there may be specialized methods. In particular, if additionally the objective function is linear in the parameters $\mathbf{x}$, we have the special case of **linear programming**. An objective that is quadratic in $\mathbf{x}$ with linear constraints gives us **quadratic programming**. The historical use of the word "programming" is unfortunate, but we will use it because it has a large literature. The term **nonlinear programming** is, as far as I can determine, a synonym for the general optimization problem of nonlinear objective with nonlinear constraints.

Equality constraints can be specified as a vector of (possibly) nonlinear functions

$$\mathbf{e}(\mathbf{x}) = \mathbf{0} \qquad (1.10)$$

We can formally replace each such equation with two inequations and thereby subsume the set of functions $\mathbf{e}$ into the expanded set of $\mathbf{c}$ functions, even if this is not the approach we would choose in seeking a solution. However, feasible equality constraints imply that we can solve for some of the parameters in terms of the rest. For the most part, I prefer to assume that such a solution has been found and that we are working with a reduced problem that does NOT have equality constraints.

1.6   Solving sets of equations

The solution of the equations that give rise to equality constraints is not quite the same as solving a complete set of n (nonlinear) equations in n unknowns. The constraint equations generally leave us with some parameters still to be determined, and general methods may be difficult to discover and apply.

For solving complete sets of equations, we want to solve for $\mathbf{x}^*$ in

$$\mathbf{e}(\mathbf{x}^*) = \mathbf{0} \qquad (1.11)$$

A general vector $\mathbf{x}$ will not give $\mathbf{0}$ but instead a non-zero vector $\mathbf{r}$, that is, "residuals." Thus we can attempt a solution using nonlinear least squares methods. If our resulting value of the objective, which is

$$f(\mathbf{x}) = \mathbf{x}' * \mathbf{x} \qquad (1.12)$$

turns out to be zero, then we have found a solution. Here I will tread very lightly on the issue of whether this solution is unique and similarly consider the question of existence of a solution to be proven by finding such a solution. A failure to find a zero sum of squares is not, however, a guarantee that there is no such solution.

In Chapter 7, we discuss nonlinear equations. As we have just noted, these could be solved by nonlinear least squares if the solution has zero residuals. This is usually thought to be a poor approach, but I have sometimes found it expedient. In the other direction, solving a set of gradient equations can solve optimization and nonlinear least squares problems if the solution can be found and can be shown to be an optimum (minimum for minimization, maximum for maximization) and not a saddle point or on a flat area. Moreover, we may want a global optimum where multiple local optima exist. Thus, I generally avoid this approach to optimization because it demands too much checking of the "solution."

1.7   Conditions for optimality

It is helpful at this early stage to mention that the meaning of "optimize" is not without interpretations. It is possible to take a very pure mathematical viewpoint but that inevitably fails to help scientists, economists, and engineers who need to provide reasonable answers to difficult questions.

For our purposes, optimality means that we want to find the values of the parameters of some function describing a system of interest that cause that function to be a minimum. To find a maximum, we can find the minimum of the negative of the function.

Beyond the concept of a minimum, we may have a whole plethora of assumptions, many of which will be unstated. These assumptions give rise to constraints, some of which could be incorporated into the optimization methods, although others may be simply of the form "that set of parameters is not of interest." This could be because there are mathematical possibilities that have no relevance to practical application, for example, solutions to a medical infection that also kills the patient.

Of particular interest in this book will be the **optimality conditions**, sometimes called the *Karush Kuhn Tucker conditions* (Karush, 1939; Kuhn and Tucker, 1951). These essentially say that, except at a constraint boundary, a minimum of a function will have zero gradient and a positive-definite Hessian. We return to these ideas several times within the book.

1.8   Other classifications

Solution methods sometimes dictate the way we approach problems. The old adage that for a man with a hammer everything looks like a nail is far too close to reality to be comfortable. Because there are many specialized computational methods that solve particular optimization-related problems very efficiently, there is often a temptation to cast problems so these methods can be applied. An important aspect of this book is to provide some perspective so that the user can decide when and if this is sensible.

References

Karush W 1939 *Minima of functions of several variables with inequalities as side constraints*. MSc Dissertation, Master's thesis, Department of Mathematics, University of Chicago, Chicago, IL.

Kuhn HW and Tucker AW 1951 Nonlinear programming. *Proceedings of 2nd Berkeley Symposium*, pp. 481–492, Berkeley, USA.

Nash JC and Walker-Smith M 1987 *Nonlinear Parameter Estimation: An Integrated System in BASIC*. Marcel Dekker, New York. See http://www.nashinfo.com/nlpe.htm for an expanded downloadable version.

2

Optimization algorithms – an overview

In this chapter we look at the panorama of methods that have been developed to try to solve the optimization problems of Chapter 1 before diving into R's particular tools for such tasks. Again, R is in the background. This chapter is an overview to try to give some structure to the subject. I recommend that all novices to optimization at least skim over this chapter to get a perspective on the subject. You will likely save yourself many hours of grief if you have a good sense of what approach is likely to suit your problem.

2.1 Methods that use the gradient

If we seek a single (local) minimum of a function $f()$, possibly subject to constraints, one of the most obvious approaches is to compute the **gradient** of the function and proceed in the reverse direction, that is, proceed "downhill." The gradient is the n-dimensional slope of the function, a concept from the differential calculus, and generally a source of anxiety for nonmathematics students.

Gradient descent is the basis of one of the oldest approaches to optimization, the **method of steepest descents** (Cauchy, 1848). Let us assume that we are at point x (which will be a vector if we have more than one parameter) where our gradient – the vector of partial derivatives of the objective with respect to each of the parameters – is $g(x)$. The method is then to proceed by a sequence of operations where we find a "lower" point at

$$x_{\text{new}} = x - \lambda g(x) \tag{2.1}$$

Nonlinear Parameter Optimization Using R Tools, First Edition. John C. Nash.
© 2014 John Wiley & Sons, Ltd. Published 2014 by John Wiley & Sons, Ltd.
Companion Website: www.wiley.com/go/nonlinear_parameter

The "minor detail" of choosing λ is central to a workable method, and in truth, the selection of a **step length** in descent methods is a serious topic that has filled many journals.

While the name sounds good, steepest descents generally is an inefficient optimization method in n parameters. A version of steepest descents was created by simplifying the **Rcgmin** package. Let us try minimizing a function that is the sum of squares

$$f(x) = \sum_{i=1}^{n} (x_i - i)^2 \tag{2.2}$$

which is expressed in R, along with its gradient as follows:

```
sq.f <- function(x) {
    nn <- length(x)
    yy <- 1:nn
    f <- sum((yy - x)^2)
    cat("Fv=", f, " at ")
    print(x)
    f
}
sq.g <- function(x) {
    nn <- length(x)
    yy <- 1:nn
    gg <- 2 * (x - yy)
}
```

This function is quite straightforward to minimize with steepest descents from most starting points. Let us use $x = (.1, .8)$. (We will discuss more about starting points later. See Section 18.2.)

```
source("supportdocs/steepdesc/steepdesc.R")
# note location of routine in directory under nlpor
x <- c(0.1, 0.8)
asqsd <- stdesc(x, sq.f, sq.g, control = list(trace = 1))

## stdescu -- J C Nash 2009 - unconstrained version CG min
## Fv= 2.25  at [1] 0.1 0.8
## Initial function value= 2.25
## Initial fn= 2.25
## 1    0   1    2.25    last decrease= NA
## Fv= 2.237  at [1] 0.1027 0.8036
## Fv= 8.4e-23  at [1] 1 2
## 3    1   2    8.4e-23    last decrease= 2.25
## Very small gradient -- gradsqr = 3.36016730417128e-22
## stdesc seems to have converged

print(asqsd)

## $par
## [1] 1 2
##
```

```
## $value
## [1] 8.4e-23
##
## $counts
## [1] 3 2
##
## $convergence
## [1] 0
##
## $message
## [1] "stdesc seems to have converged"
```

However, a nonparabolic function is more difficult. From the same start, let us try the well-known Rosenbrock function, here stated in a generalized form. We turn off the trace, as the steepest descent method uses many function evaluations.

```
# ls() 1 to (n-1) variant of generalized rosenbrock function
grose.f <- function(x, gs = 100) {
    n <- length(x)
    1 + sum(gs * (x[1:(n - 1)] - x[2:n]^2)^2 + (x[1:(n - 1)] - 1)^2)
}

grose.g <- function(x, gs = 100) {
    # gradient of 1 to (n-1) variant of generalized rosenbrock function
    # vectorized by JN 090409
    n <- length(x)
    gg <- as.vector(rep(0, n))
    tn <- 2:n
    tn1 <- tn - 1
    z1 <- x[tn1] - x[tn]^2
    z2 <- x[tn1] - 1
    gg[tn1] <- 2 * (z2 + gs * z1)
    gg[tn] <- gg[tn] - 4 * gs * z1 * x[tn]
    gg
}
x <- c(0.1, 0.8)
arksd <- stdesc(x, grose.f, grose.g, control = list(trace = 0))
print(arksd)

## $par
## [1] 0.9183 0.9583
##
## $value
## [1] 1.007
##
## $counts
## [1] 502 199
##
## $convergence
## [1] 1
##
## $message
## [1] "Too many function evaluations (> 500) "
```

We will later see that we can do much better on this problem.

2.2   Newton-like methods

Equally if not more historic is Newton's method. The original focus was on finding the roots of functions, and the optimization version attempts to provide the step toward the minimum by approximately solving for the point at which the gradient will be zero. That is, we do not directly optimize but try to find the location of the valley bottom where there is no slope. There is some debate over where and when the method was actually published, and I will not add to this here. Richard Anstee gives a useful historical discussion at the end of his tutorial www.math.ubc.ca/~anstee/math184/184newtonmethod.pdf.

Certainly, anything similar to a modern view of the method was not used by Newton or Raphson, whose names are often linked to it.

The modern view of the method is as follows. In one dimension, we start at point x and compute the gradient $f'(x)$ and second derivative $f''(x)$. We then solve for

$$\delta = -f'(x)/f''(x) \tag{2.3}$$

and repeat the process from

$$x_{new} = x + \delta \tag{2.4}$$

Let us try this from $x = 1$ with the function

$$f(x) = \exp(-.05 * x) * (x - 4)^2 \tag{2.5}$$

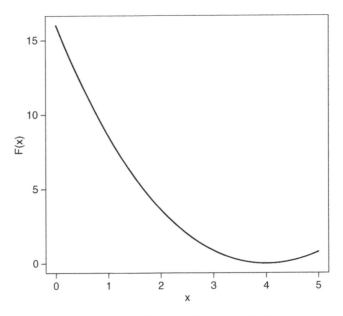

Figure 2.1   Graph of Equation (2.5).

```
F <- function(x) exp(-0.05 * x) * (x - 4)^2
curve(F, from = 0, to = 5)

newt <- function(x) {
    xnew <- x - (exp(-0.05 * x) * (2 * (x - 4)) - exp(-0.05 * x) * 0.05
        * (x - 4)^2)/(exp(-0.05 * x) * 2 - exp(-0.05 * x) * 0.05 *
        (2 * (x - 4)) - (exp(-0.05 * x) * 0.05 * (2 * (x - 4))
        - exp(-0.05 * x) * 0.05 * 0.05 * (x - 4)^2))
}
x <- 1
xold <- 0
while (xold != x) {
    xold <- x
    cat("f(", x, ")=", F(x), "\n")
    x <- newt(xold)
}

## f( 1 )= 8.561
## f( 3.459 )= 0.2457
## f( 3.979 )= 0.0003603
## f( 4 )= 8.871e-10
## f( 4 )= 5.407e-21
## f( 4 )= 0
```

A **warning**: If you read articles about Newton's method that view it as root-finding, say of function $q(x)$, then the iteration equation is in the form of

$$\delta = -q(x)/q'(x) \tag{2.6}$$

Thus the optimization view requires **two** degrees of derivatives of the original function $f(x)$. This is a serious amount of mental effort in most problems.

The n-parameter Newton method uses the Hessian matrix of second derivatives of the objective function with respect to the parameters, which is the generalization of the second derivative for the 1D problem above. Starting at some set of parameters **x**, we compute **H** and **g**. We then solve

$$\mathbf{H} * \mathbf{delta} = -\mathbf{g} \tag{2.7}$$

and try to compute the objective $f(\mathbf{x}_{new}) = f(\mathbf{x} + \mathbf{delta})$

2.3   The promise of Newton's method

Newton's method is attractive because there are theoretical results that show it is extremely efficient under some conditions that are not too unlikely. Principally, the quality of Newton's method that is attractive is **quadratic convergence**, which means that if we are "close enough" to a solution x_S with our starting estimate x_0, then after some iterations, the distances to the solution will obey

$$\lim_{k \to \infty} (x_{k+1} - x_S)/(x_k - x_S)^2 \tag{2.8}$$

A statement that is easier to understand is that the number of correct digits approximately doubles with each iteration. This is more or less what is seen in the earlier example, in which convergence is very rapid.

2.4   Caution: convergence versus termination

This is a good point to make the first of what will be several mentions that the theoretical concept of the **convergence** of iterative algorithms can be very important for understanding them, but it may be quite unlike the actual **termination** of the program implementing that algorithm.

Algorithms converge; programs terminate.

We generally need to put in very carefully considered tests and conditions to ensure that we do, in fact, terminate (Nash, 1987).

It is useful to be a bit persnickety and insist on keeping "convergence" for algorithms or methods and "termination" for their implementations because this will remind us that the programs have to deal with situations that are outside the purview of the mathematical niceties. For example, it is very easy in this era of "cut and paste" to get character formatting information into what looks like a simple number, giving some sort of error as our optimization code misinterprets the information we are supplying.

2.5   Difficulties with Newton's method

There are lots of dangerous beasts in the night-time optimization forest. Most of these can be seen in one-parameter problems, but it is worth noting that n dimensions usually make things a lot more difficult, because the issues have essentially an n-fold possibility of occurrence. The Wikipedia article http://en.wikipedia.org/wiki/Newton's_method presents these issues very clearly, along with a lot of other well-prepared material. These difficulties can be overcome with careful attention to detail. We list the main issues here, illustrated in one dimension:

- singular Hessian, for example, $f(x) = x$;

- multiple minima, for example, $f(x) = \sin(x)$;

- constraints, for example, $f(x) = x^2$ subject to $x > 3$;

- undefined function, for example, $f(x) = \log(x)$ when x approaches 0;

- too big a step, for example, $f(x) = \sin(x)$ with a step bigger than $2 * \pi$ which will move the parameter to the next trigonometric cycle.

2.6   Least squares: Gauss–Newton methods

Statisticians often deal with problems that involve objective functions that are sums of squares. The classic linear regression and analysis of variance tools of multivariate statistics minimize sums of squared residuals, although most users of the lm() linear modeling function in R would not necessarily have that viewpoint. Moreover, there are a number of situations in which the traditional problem is modified so that either the parameter estimates or other quantities are constrained. In such cases, other tools are needed. Sometimes an optimization tool can be a practical approach to finding a solution.

As with its linear counterpart, the **nonlinear least squares** problem is so common that special methods have been developed and should be considered when such problems are to be solved. There is a huge literature, and many factional disputes on how things should be done.

The main approach – and one that is used in R's nls() function – is the Gauss–Newton method. Our objective function $f(\mathbf{x})$ is the sum of nobs squared elements that we may consider as **residuals**. If we have a model $m(\mathbf{x}, \text{data})$ that involves our parameters $\mathbf{x}$ and some other data to model a variable $\mathbf{y}$ (here explicitly written as a vector of observations), we can write the nobs residuals as

$$\mathbf{r} = m(\mathbf{x}, \text{data}) - \mathbf{y} \tag{2.9}$$

This is the "wrong way round" for most statisticians, including Gauss, but because we are squaring the residuals, the objective function is the same. I like this way because the derivatives of the residuals and the model then have the same sign, and sign errors are so common that I do everything I can to avoid them. If you are happier with the conventional "data-model" formulation, please feel free to use it. This is a matter of preference, not law.

If we try to apply the Newton method to a nonlinear least squares problem, we see that the gradient of the objective function is

$$\mathbf{g} = J'\mathbf{r} \tag{2.10}$$

where

$$J_{i,j} = \partial r_i / \partial x_j \tag{2.11}$$

gives the i, j element of the **Jacobian** of the residuals with respect to $\mathbf{x}$.

However, the Hessian is now more complicated,

$$H_{j,k} = J' \cdot J_{j,k} + \sum_{i=1}^{\text{nobs}} r_i \cdot (\partial^2 r_i / \partial x_j \partial x_k) \tag{2.12}$$

The second part of this expression involving the second partial derivatives of the residuals is generally "ignored" on the basis that the residuals are usually "small." This is, of course, an interesting twist on the argument that nls is not useful for small residual problems. The practical reason is that these derivatives require us to compute nobs square matrices each of a size equal to the number of parameters – a big chore.

A very successful "cheat" is to approximate the second term with some other matrix. Donald Marquardt (Marquardt, 1963) developed one of the most successful early nonlinear least squares codes by adding a scaled unit matrix to the Jacobian inner product, that is,

$$\text{Happ} = J' \cdot J + \lambda * I \tag{2.13}$$

The effect of this is that when λ is very large, we solve

$$I * \delta = -\mathbf{g} \tag{2.14}$$

and get a steepest descents search direction. When λ is small, we have the Gauss–Newton search direction. In practice, with a strategy that increases λ when we fail to reduce the sum of squares and decreases it when we succeed, this works very well. Of course, there are numerous discussions on the tactical details of the settings for increase, decrease, and initial λ. As far as I can determine, sometime after Marquardt's paper appeared, it was noted that the ideas had been suggested sometime during World War II by Levenberg (1944), and his name is sometimes added to the method.

Marquardt also showed that using a matrix D instead of I, defined

$$D_{i,j} = \begin{cases} (J' \cdot J)_{i,i} & \text{if} \quad i = j \\ 0 & \text{if} \quad i <> j \end{cases} \tag{2.15}$$

avoids some issues when the parameters are of widely different scale, as in the Hobbs weeds problem (see Section 1.2). However, when I came to that problem, I was working (in 1974) on a Data General NOVA in BASIC, where the floating point arithmetic used six hexadecimal digits without a guard digit. In computing the sum of squares and cross products $J' \cdot J$, some of the diagonal elements underflowed to zero, which rendered the stabilization ineffective. I found that the unscaled version (using just an identity matrix I) worked, but even better was a combination

$$D_{\text{modified}} = D + \phi * I \tag{2.16}$$

where ϕ is a smallish number (I generally use 1).

Other approaches abound, such as using a line search with the original Gauss–Newton method (Hartley, 1961) or solving the Gauss–Newton equations with a singular value decomposition and dropping directions that correspond to small singular values (Jones, 1970). My experience is that the Marquardt approach with attention to the possibility of underflow works as well as any other method and has the merit of simplicity.

2.7   Quasi-Newton or variable metric method

We have seen that Newton's method obtains a search direction from a solution of the equation

$$H\delta = -g \tag{2.17}$$

From this, many stabilized versions of Newton's method can be created using different strategies and tactics to search along δ and control for potential mishaps. However, the fundamental chore in Newton's method is computing the Hessian H. Even the gradient g is a considerable difficulty, but for the moment, we will take this as feasible.

As with the Gauss–Newton method, we will try to approximate H in ways that are easier to compute. It can be argued that the most effective approach is the family of algorithms called quasi-Newton methods. These begin with an approximation either to H or its inverse H_{inverse}. The following are the steps of any of the quasi-Newton method, with some variants referred to as *variable metric methods*:

- Provide starting parameters x_0, set an initial "approximate" Hessian or Hessian inverse, and compute the initial function $f0$ and gradient g, as well as any settings needed to control a particular method.

- "Solve" the Newton equations to get a search direction δ. For methods using an approximate Hessian inverse, this is simply a matrix multiplication. Other approaches may solve the equations explicitly, or else work with decompositions of the approximate Hessian or approximate Hessian inverse.

- Perform some kind of line search along the search direction δ. A failure to improve the function value is one component of the termination tests.

- Update the approximate Hessian or approximate Hessian inverse in some way. There are a number of approaches, but most are based on two families of update formulas, one labeled "Davidon Fletcher Powell" (DFP) and the other "Broyden–Fletcher–Goldfarb Shanno" (BFGS). However, there are many details that may be important.

- Apply various termination tests and possibly restart the process if progress is not being made.

To illustrate this, the "BFGS" method of `optim()` uses an identity matrix as the initial approximate inverse Hessian, so the first search direction is steepest descents. The line search simply shortens the step along this direction until a "satisfactory decrease" is found. This is sometimes called backtracking to an acceptable point. The BFGS formula is used to update the inverse Hessian approximation. If the line search is not successful, the inverse Hessian approximation is reset to the identity and we only terminate when the line search fails for this steepest descents direction.

The unconstrained optimizer Rvmminu from package **Rvmmin** is supposed to be the same algorithm, but I note that the sequence of evaluation points for some trial problems is not the same with this code and with optim 'BFGS'. However, very small changes in arithmetic or control settings can alter trajectories of minimizers, so this is not totally surprising.

By comparison, the minimizer **ucminf** uses much the same approach but with a quadratic inverse interpolation line search (Nielsen, 2000). This appears to be slightly more efficient than optim method 'BFGS', and I have toyed with trying to implement it. However, the code and description are sufficiently complicated that I have not done so. My motive for reimplementing the method was to add bounds constraints (Chapter 11).

2.8   Conjugate gradient and related methods

A somewhat different approach to gradient-based function minimization is the family of nonlinear conjugate gradient minimizers. These are derived as if one is minimizing a quadratic form with the Hessian a constant positive definite matrix. This leads to an algorithm that has the following structure:

- Provide starting parameters $\mathbf{x}_0$, and compute the initial function $f0$ and gradient $\mathbf{g}$. Choose the initial search direction to be the negative gradient.

- Perform some kind of line search along the search direction. A failure to improve the function value stops the method if we are searching along the negative gradient, or else we reset the search to this direction.

- We compute the gradient at the new point and use it, the previous gradient and the search direction to derive a new search direction. Note that this involves only a few vectors of storage. There are several formulas to update the search direction, and a number of tests to ensure that this is a "good" direction.

- Apply various termination or restart tests and continue until progress is not being made.

Despite the very low storage requirement and rather simple structure of CG methods, they often perform surprisingly well on quite large problems (i.e., in hundreds of parameters). On the other hand, "often" is not the same as "almost always," and it is important to note that I regard the 'CG' method in optim() and its precursor in Nash (1979) as the least successful of the methods I implemented. This offers a choice of three formulas for updating the search direction, namely, the original Fletcher–Reeves formula, one due to Polak and Ribière, or one due to Beale and Sorenson. The first of these is the default in optim(); however, the second is my reluctant preference.

Fortunately, we need not fuss with such choices, Dai and Yuan (2001) showed a way to combine the formulas and use them to provide a test of when to restart the iteration. This is implemented in package **Rcgmin** and appears to work well.

There is more recent work by Hager and Zhang (2006) that may offer further improvement, but I have not had an opportunity to implement these ideas.

2.9   Other gradient methods

For a problem in n parameters, a quasi-Newton method requires working storage for a matrix of order n plus several vectors of the same order. A conjugate gradients method is much more space efficient – just a few n vectors – but may not be as good at minimizing the function in a few evaluations. Thus many researchers have sought to find ways to get both time and space efficiencies by methods that are somehow intermediate between the two families. One of these is represented in optim, the Limited Memory BFGS method of Nocedal and colleagues (Zhu et al. 1997). Another is the inexact Newton method and truncated-Newton method (Nash, 2000b), where we solve the (linear) Newton equations partially, in particular for the truncated-Newton methods by approximate iterative methods such as conjugate gradients. There are many details in such methods that are sometimes critical for efficiency in finding the solution. Worse, in my experience, these details can result in very large improvements in the methods for some problems but be detrimental in others.

2.10   Derivative-free methods

When analytic derivatives are extremely difficult or impossible to provide, it is necessary to use methods that only depend on the function value. I would echo the sentiments expressed in Conn et al. (2009) that methods for derivative-free optimization are a necessity, but they should not be used if gradients are available.

2.10.1   Numerical approximation of gradients

For problems where the number of parameters is small and the underlying function has derivatives (but we are unwilling or unable to generate the code for them), numerical approximation of the gradient is sometimes a feasible option. Often very crude approximations are adequate to get a reasonable solution, but possibly inefficiently. The use of numerical derivative approximations is discussed in Chapter 10.

2.10.2   Approximate and descend

An alternative approach to working at a single point in the domain of the parameters is to choose a set of points and model the objective function from the values at those points. The model needs only approximate the objective function crudely, although we generally assume that at the minimum it is "close enough." Once we have a model, we compute the minimum (or some lower point) of the modeling function and move to its minimum. If the objective function at that point is "better" in some sense than points already evaluated, the new point is added to the modeling set, one or more other points may be dropped, and the process repeated. Such an "approximate and descend" approach can also be used when the objective function can only be approximately evaluated, although the uncertainty in the evaluation then implies that we must use more approximating points. See Nash (2003) for one such approach.

2.10.3   Heuristic search

A brute-force approach is to simply evaluate the objective function at a variety of points and then repeat the process near the lowest point or points. Either grid or random search can, however, be improved by searching more cleverly. There are many ways to do this, and they use different ideas and assumptions about how to search for a "better" point.

An early approach, using a simple search along the parameter axes (axial search) followed by a pattern move, with repetition, was suggested by Hooke and Jeeves (1961). This tends to be quite reliable but inefficient. However, Torczon (1995) has extended and improved these pattern search ideas.

A different, and popular, approach for a function of n parameters, starts with $n + 1$ points in a polytope (or simplex) that has the points not in a hyperplane. Various schemes were developed for evolving the polytope to find the minimum. The most used seems to be that of Nelder and Mead 1965, which uses the reflection of the highest point through the centroid (average) of the rest as the basis of a set of tactical adjustments to the polytope (Nash, 1979, Chapter 14). Despite many papers about the deficiencies of the Nelder–Mead method, it remains one of the most used, and all but a few of the "improvements" to it have been abandoned. It is, however, worth mentioning the ideas of Kelley (1997) that have been implemented in functions nmk() and nmkb() of package **dfoptim**.

2.11   Stochastic methods

When the objective function has many minima, and the global minimum is sought, stochastic methods are often attempted. There is a plethora of such methods. So many, in fact, that I suspect there may be several named methods that are actually the same algorithms.

Some of the many names are

- Simulated Annealing (Kirkpatrick et al. 1983)

- Taboo Search (Glover, 1989)

- Quantum annealing
  (http://en.wikipedia.org/wiki/Quantum_annealing)

- Probability Collectives (Rajnarayan et al. 2006)

- Reactive search optimization (http://www.reactive-search.org)

- Cross-entropy method (Boer et al. 2002)

- Random search, of which there are a number of approaches, one of the earliest by Bremmerman (1970)

- Stochastic tunneling (Wenzel and Hamacher, 1999)

- Parallel tempering, also known as *replica exchange MCMC sampling*, (Geyer, 1991; Swendsen and Wang, 1986)

- Stochastic hill climbing or stochastic gradient descent (various workers appear to use these terms in conjunction with machine learning and genetic algorithms)

- (Particle) swarm algorithms (Kennedy and Eberhart, 1995)

- Evolutionary operation, begun in the 1950s by George Box for process optimization, which has influenced the development of other evolutionary algorithms

- Genetic algorithms, which are one form of evolutionary algorithms, with many contributions starting as early as the 1950s, but possibly established by Holland (1975).

In addition, of course, there are many user-developed approaches to stochastic optimization. If the number of local optima is small, then random starts to conventional minimizers is a common and often-effective strategy. For example, Varadhan and Gilbert (2009a) provide function `multiStart` in package **BB**.

The major source of dissatisfaction with stochastic approaches to optimization is that they lack any guarantee that they have succeeded. Truthfully, methods for global optimization with guarantees of success are elusive or difficult to use or both, otherwise they would have replaced the stochastic methods. For particular functional forms, we may be able to apply specialized methods such as Hompack (Watson et al. 1997) or interval analysis (Hansen and Walster, 2004), but most users will have difficulty in confirming that their problems fit the conditions required for such methods. There are also, to my knowledge, no packages for use of such techniques in R.

2.12   Constraint-based methods – mathematical programming

While R is weak in mathematical programming (MP) tools, it is important to know that there is a huge family of tools in the wider community for solving optimization problems that are primarily defined through the constraints that must be imposed on the parameters. The linear programming (LP) problem, where we wish to minimize a linear function of the parameters subject to a number of inequality constraints, often combined with nonnegativity conditions on the parameters, is one of the classical computational problems with a very large literature. The original Dantzig Simplex method is still a subject of interest (Nash, 2000).

When we have a quadratic objective function, but still linear inequalities, we have the quadratic programming (QP) problem. Linearly constrained least squares problems can, therefore, be solved by QP methods. More general MP problems, sometimes called *nonlinear programming* (NLP), can be very difficult to properly formulate and solve. They may have no solution, or multiple solutions, or the solver may fail to find an acceptable solution. However, methods are gradually being developed for some classes of problems of this type.

Convex optimization, where the objective and constraints are convex functions over the parameter space and the set of allowed parameters is also a closed convex set, provides a class of optimization problems that is at the same time quite useful yet without tools in R that are designed for their solution.

R also tends to deal much more with continuous (i.e., 'real') numbers, while many mathematical programming tasks concern choices, assignments, sequences, and other essentially discrete parameters that are handled better by integers.

References

Boer PTD, Kroese D, Mannor S and Rubinstein R 2002 A tutorial on the cross-entropy method. Annals of Operations Research 134, 19–67.

Bremmerman H 1970 A method for unconstrained global optimization. Mathematical Biosciences 9, 1–15.

Cauchy A 1848 Mthode gnrale pour la resolution des systmes dquations simultanes. Comptes Rendus de l'Acadmie des Sciences 27, 536–538.

Conn AR, Scheinberg K and Vicente LN 2009 Introduction to Derivative-Free Optimization. Society for Industrial and Applied Mathematics, Philadelphia, PA.

Dai YH and Yuan Y 2001 An efficient hybrid conjugate gradient method for unconstrained optimization. Annals of Operations Research 103(1–4), 33–47.

Geyer CJ 1991 Markov chain Monte Carlo maximum likelihood Computing Science and Statistics, Proceedings of the 23rd Symposium on the Interface, p. n.a.. American Statistical Association, New York.

Glover F 1989 Tabu search - part 1. ORSA Journal on Computing 1(2), 190–206.

Hager WW and Zhang H 2006 Algorithm 851: CG_DESCENT, a conjugate gradient method with guaranteed descent. ACM Transactions on Mathematical Software 32(1), 113–137.

Hansen ER and Walster GW 2004 Global Optimization Using Interval Analysis. MIT Press, Cambridge, MA.

Hartley HO 1961 The modified Gauss–Newton method for the fitting of non-linear regression functions by least squares. Technometrics 3, 269–280.

Holland JH 1975 Adaptation in Natural and Artificial Systems. University of Michigan Press, Ann Arbor, MI.

Hooke R and Jeeves TA 1961 "direct search" solution of numerical and statistical problems. Journal of the ACM 8(2), 212–229.

Joe H and Nash JC 2003 Numerical optimization and surface estimation with imprecise function evaluations. Statistics and Computing 13(3), 277–286.

Jones A 1970 Spiral – a new algorithm for non-linear parameter estimation using least squares. Computer Journal 13(3), 301–308.

Kelley CT 1997 Detection and remediation of stagnation in the Nelder–Mead algorithm using a sufficient decrease condition. SIAM Journal on Optimization 10, 43–55.

Kennedy J and Eberhart R 1995 Particle swarm optimization. Proceedings of IEEE International Conference on Neural Networks, 1995, Vol. 4, pp. 1942–1948.

Kirkpatrick S, Gelatt CD Jr. and Vecchi MP 1983 Optimization by simulated annealing. Science 220(4598), 671–680.

Levenberg K 1944 A method for the solution of certain non-linear problems in least squares. Quarterly of Applied Mathematics 2, 164–168.

Marquardt DW 1963 An algorithm for least-squares estimation of nonlinear parameters. SIAM Journal on Applied Mathematics 11(2), 431–441.

Nash JC 1979 Compact Numerical Methods for Computers: Linear Algebra and Function Minimisation. Adam Hilger, Bristol. Second Edition, 1990, Bristol: Institute of Physics Publications.

Nash JC 1987 Termination strategies for nonlinear parameter determination. In Proceedings of the Australian Society for Operations Research Annual Conference (ed. Kumar S), pp. 322–334, Melbourne, Australia.

Nash JC 2000a The Dantzig simplex method for linear programming. Computing in Science and Engineering 2(1), 29–31.

Nash SG 2000b A survey of truncated-Newton methods. Journal of Computational and Applied Mathematics 124, 45–59.

Nelder JA and Mead R 1965 A simplex method for function minimization. Computer Journal 7(4), 308–313.

Nielsen HB 2000 UCMINF - an algorithm for unconstrained, nonlinear optimization. Technical report, Department of Mathematical Modelling, Technical University of Denmark. Report IMM-REP-2000-18.

Rajnarayan D, Wolpert D and Kroo I 2006 Optimization under uncertainty using probability collectives. Proceedings of 11th AIAA/ISSMO Multidisciplinary Analysis and Optimization Conference, Portsmouth, VA. AIAA-2006-7033.

Swendsen RH and Wang JS 1986 Replica Monte Carlo simulation of spin glasses. Physical Review Letters 57, 2607–2609.

Torczon V 1995 Pattern search methods for nonlinear optimization. SIAG/OPT Views and News 6, 7–11.

Varadhan R and Gilbert P 2009 BB: an R package for solving a large system of nonlinear equations and for optimizing a high-dimensional nonlinear objective function. Journal of Statistical Software 32(4), 1–26.

Watson LT, Sosonkina M, Melville RC, Morgan AP and Walker HF 1997 Algorithm 777: Hompack90: a suite of fortran 90 codes for globally convergent homotopy algorithms. ACM Transactions on Mathematical Software 23(4), 514–549.

Wenzel W and Hamacher K 1999 Stochastic tunneling approach for global minimization of complex potential energy landscapes. Physical Review Letters 82, 3003–3007.

Zhu C, Byrd RH, Lu P and Nocedal J 1997 Algorithm 778: L-bfgs-b: fortran subroutines for large-scale bound-constrained optimization. ACM Transactions on Mathematical Software 23(4), 550–560.

3

Software structure and interfaces

Having presented some of the problems that fall under the umbrella that is the title of this book, we now turn to consider different ways to structure software to try to find solutions to those problems. For discussion purposes and because it is the class of problems with which I have the most experience, we will primarily focus on function minimization with at most quite simple constraints. In the statistical community, many people call this **optimization**, while those from engineering and economics would expect problems with many more constraints but a possibly simpler objective function. As R is largely aimed at statistical computations, the treatment here will lean in that direction.

3.1 Perspective

In a fairly long career in this area, I have found that there is a common love of "speed." More correctly, people do not want to feel that they are waiting for their calculations to finish. If they do find that they are waiting, they may begin to judge methods on the basis of timings, sometimes surprisingly few and poorly conducted timings.

While sharing a strong interest in computational efficiency, I take the following views.

- It is important to know that the answer presented is "right," or at least to provide guidance or information if the result may not be satisfactory. For example, there may be multiple solutions, or the result may be simply the best we can do in a difficult situation. Optimizers can and do stop for reasons

Nonlinear Parameter Optimization Using R Tools, First Edition. John C. Nash.
© 2014 John Wiley & Sons, Ltd. Published 2014 by John Wiley & Sons, Ltd.
Companion Website: www.wiley.com/go/nonlinear_parameter

other than that there is a satisfactory answer. User functions frequently have errors, despite our best efforts. Uncovering and correcting wrong answers can be very very time consuming for humans, as well as being potentially embarrassing.

- Doing the tests and checks required to provide the diagnostic and explanatory information just mentioned will take computational time. Hence, users will find, for example, that running the optimizers in the package **optimx** (see Section 9.1) takes longer than running them directly. The penalty generally is not huge, although computing Hessians when the number of parameters n is large can take longer than the optimization process. Indeed, **optimx** does not compute the Hessian and optimality (Karush–Kuhn–Tucker, KKT) tests when $n > 50$ if there are no analytic derivatives available or $n > 500$ when there are. Similarly, the experimental package **optplus** provides parameter scaling and checking for inadmissible parameters on every function evaluation, but the cost is high enough that I am rethinking how to better provide such features.

- When a problem must be solved very quickly and/or very frequently, so that the time taken by R optimizers is too great, then I believe it is worth considering reimplementing the particular calculation. See Chapter 19. Otherwise, I feel that computer time is "cheap" relative to that of the user.

3.2   Issues of choice

While this book is about optimization in R, more generally users must decide on the system under which they will perform computations. In the past decade, it has become possible to decide to compute in "R" without reference to the hardware or operating system to be used. This is largely true of other major computational choices, such as Matlab or Python or SAS or etc.

Having decided on R, however, there are so many packages – well above 4000 at the time of writing – that it is difficult to discover what each of them does and impossible to decide on whether their features and quality make them suitable to our needs. This is especially difficult for optimization problems. Many packages are targeted at specific domains of research. They may claim great performance or features, but those capabilities may be applicable to a minuscule collection of users.

Clearly, one of the purposes of this book is to provide guidance through this undergrowth, but even though I am active in pursuing the subject of optimization in R, I am painfully conscious that I am going to miss some package or other with a useful capability. Therefore, I have chosen to look at the general tools that can be adapted to many problems rather than tools for specific problems.

It is also notable that there are many packages that include custom-built optimizers, for example, package **likelihood** (Murphy, 2012). From the perspective of a tool builder like myself, this is unfortunate:

- It requires generally quite a lot of work to abstract the optimizer from the surrounding infrastructure and context in such packages.

- The interests and focus of the package developer are generally on the particular problems of their domain of work. The optimizer may be well constructed for that domain but risky to use in other areas.

- For similar reasons, the provenance and documentation of the optimizer may be unclear.

Moreover, I am unapologetic about favoring packages that are written entirely in R and which present themselves as simply and generally as possible. This focus is partly due to a level of laziness – I do not enjoy digging through difficult and opaque documentation with idiosyncratic vocabulary. However, I am also convinced that there are far too many situations where using programs written in languages other than R gives rise to errors in interfacing or leaves unsuitable controls in place in the non-R code. Furthermore, it is difficult to pin down the provenance of the codes, even though R does very well in documenting its origins.

For example, in the case of the optimizer L-BFGS-B, a search through the source package yields the routine lbfgs.c that has a comment that it is 'version 2.3'. Yet the web page of the original authors (http://www.ece. ~northwestern.edu/nocedal/lbfgsb.html) shows only versions 2.1 (1997) and 3.0 (2011). Moreover, both these are in **Fortran** not C. There are notes in the R source in various places that code has been converted from **Fortran** to C using a program called f2c (Feldman et al. 1991). It seems likely that a partially updated version of the Fortran code was converted to C in the late 1990s.

I can sympathize with the difficulty of choosing, converting, and interfacing a complicated program such as L-BFGS-B for use in R. Further, I can appreciate why developers would prefer to not have to reimplement the method in R itself. However, I have found that once a code is in R, and especially if the coding is kept plain – I often use the adjective pedestrian – it is much much easier to find and fix bugs, to adjust the code to particular situations, or to improve it as new ideas become current.

3.3   Software issues

There are rather a lot of software issues that arise in connection with attempted computational solution of optimization problems:

- Specifying the objective and constraints to the optimizer

- Communicating exogenous data to problem definition functions

- Masked (i.e., temporarily fixed) optimization parameters

- Dealing with inadmissible results

- Providing derivatives for functions

- Derivative approximations when there are constraints

- Scaling of parameters and function

- Termination tests – normal finish to computations

- Termination tests – abnormal ending

- Output to monitor progress of calculations

- Output of the optimization results

- Controls for the optimizer

- Default control settings

- Appropriate measures of performance

- The optimization interface.

We will explore these issues in the rest of the chapter.

3.4   Specifying the objective and constraints to the optimizer

Unlike statistical computations with a well-established structure, for example, linear regression, the general constrained optimization problem requires the user to supply an objective function $f(\mathbf{x})$ as well as any inequality constraints

$$\mathbf{c}(\mathbf{x}) >= \mathbf{0} \tag{3.1}$$

or equality constraints

$$\mathbf{e}(\mathbf{x}) = \mathbf{0} \tag{3.2}$$

The wide variety of possible functional forms leads to problems of very diverse levels of difficulty.

The latter, equality, constraints may, of course, be introduced in other ways, ways that may unfortunately destabilize the computations. Overall, this issue is **specifying the objective and constraints to the optimizer**.

3.5   Communicating exogenous data to problem definition functions

In the real world, the objective function $f()$ and the constraints $\mathbf{c}$ and $\mathbf{e}$ are not only functions of the parameters $\mathbf{x}$ but also depend on data. In fact, they may depend on vast arrays of data, particularly in statistical problems involving large systems.

Typically, such data consists of *M* observations on a number of variables. *M* can be a very large number. Communicating this data to the objective and constraint functions as well as to the functions to compute their derivatives is a sometimes awkward task. Even when it is straightforward, it can be tedious and error prone. This difficulty is magnified when one wishes to apply a number of different minimization algorithms, because each may have its own structure.

In R, data available in the user workspace is generally available to lower level routines. We can see this in the following example, where xx and yy do not need to be passed in the call to nlsLM() of package **minpack.lm**. On the other hand, function nlxb() of package **nlmrt** requires us to specify a data frame data as an argument in its call.

```
xx <- 1:12
yy <- exp(-0.1 * sin(0.3 * xx))
require(minpack.lm, quietly = TRUE)
strt1 <- list(p1 = 0, p2 = 1)
## here we do not specify any exogenous data
anls1 <- nlsLM(yy ~ exp(p1 * sin(p2 * xx)), start = strt1, trace = FALSE)
anls1

## Nonlinear regression model
##   model: yy ~ exp(p1 * sin(p2 * xx))
##    data: parent.frame()
##       p1       p2
## -0.0113   0.9643
##   residual sum-of-squares: 0.0496
##
## Number of iterations to convergence: 7
## Achieved convergence tolerance: 1.49e-08

rss <- function(par) {
    p1 <- par[1]
    p2 <- par[2]
    res <- exp(p1 * sin(p2 * xx)) - yy
    sum(res * res)
}
## check the initial sum of squares
print(rss(c(-0.1, 0.3)))

## [1] 0

## and the final sum of squares
print(rss(coef(anls1)))

## [1] 0.04957

# Now try with an optimizer (default method=Nelder-Mead)
anm <- optim(strt1, rss)
anm

## $par
##       p1       p2
## -0.01131  0.96426
##
## $value
## [1] 0.04957
```

```
##
## $counts
## function gradient
##       53       NA
##
## $convergence
## [1] 0
##
## $message
## NULL

## But (at time of writing) nlmrt requires explicit data
mydat <- data.frame(xx = xx, yy = yy)
require(nlmrt, quietly = TRUE)
anlxb1 <- nlxb(yy ~ exp(p1 * sin(p2 * xx)), start = strt1, trace = FALSE, data
= mydat)
## Insert following into call to get a more aggressive search
## control=list(roffset=FALSE, smallsstest=FALSE)
print(anlxb1)

## nlmrt class object: x
## residual sumsquares =  0.049565  on  12 observations
##     after 12   Jacobian and  20 function evaluations
## name     coeff       SE       tstat     pval      gradient    JSingval
## p1     -0.0113074  0.02762   -0.4094   0.6909    -2.513e-10   2.552
## p2      0.964283   0.3681    2.62      0.02561   9.805e-09    0.1913
```

R offers a mechanism in function calls for the provision of data that is to be passed through a function but is not specifically declared for that function. This is the ' ... ' or dot-dot-dot argument(s) to a function. That is, the called function optimizer will be declared as

```
optimizer<-function(parameters, ufn, ...)
   (code)
```

while the calling script will be written something like

```
myanswer<-optimizer(mypar, objective, datax=Xdata, ydata=Y)
```

R conveniently provides the datax and ydata information to functions within optimizer. Furthermore, within optimizer, we can write

```
fval<-ufn(parameters, ...)
```

to evaluate the function with parameters that are initially at values given by mypar and the exogenous data datax and ydata. This assumes that the data have not been altered by optimizer. The parameters clearly will be altered from the initial values mypar to try to find lower function values.

The dot argument mechanism, while fairly straightforward, can still allow of mistakes. Moreover, we need to carry quite a lot of information through the optimizer and possibly subsidiary routines such as those for scaling and numerical derivative approximation. It is, in fact, at the level of these subsidiary routines that the dots mechanism can become cumbersome. However, we can **sometimes** simplify the process somewhat by making the user data (in our example `datax` and `ydata`) essentially global to `optimizer`.

The four approaches (possibly there are others) to passing data into functions are thus

- using objects that are in the user's workspace;

- passing data via a named argument as in `nlxb()` above;

- passing objects through the `dots` mechanism;

- creating a global structure (see Section 3.5.1).

I have generally found that I get into less difficulties if I ensure that I explicitly specify the exogenous data, either with a named object or data frame that is required by the called function, or by named items passed via the `dots` mechanism.

3.5.1   Use of "global" data and variables

By and large, R scoping works fairly well to allow R functions to use objects the user has defined in his or her workspace. Unfortunately, "fairly well" can still get us into lots of trouble when there are objects used in the function that have the same name as in the workspace. In such cases, we must be careful to use the appropriate assignments in calls to the functions, such as

```
result <- afunction(x=myx, y=y)
```

This is not difficult, and I recommend using the explicit assignment rather than trusting that there will be no problem. I suspect that I am not alone in finding these scoping matters more troublesome than I would like.

Unfortunately, using the explicit syntax can become tedious within a collection of functions where we need to pass the same information around and especially where we modify it. In my own work, this affects calls to line searches within optimizers, for example, where we would like to have access to various quantities already computed such as current and previous gradients and function values. These objects are present in the scope of the main optimization routine, but they may be wanted in the line search. Moreover, the line search may update the "best" function value and its gradient, needing to then adjust the previous ones. Passing and returning all this information is work for the programmer. In R it can also imply that copies of information are made unnecessarily.

An alternative, which is sometimes useful, is to use a set of global data. This is usually discouraged by R gurus. However, it has a fairly long tradition in scientific computing, for example, in the named COMMON blocks of Fortran. We can use something similar in R by defining a list and making it into what R terms an environment.

```
mystart <- function() {
    # JN: Define globals here.
    gtn <- list(x = 0, y = 1, vec = rep(0, 9))
    envjn <- list2env(gtn)
}
y <- 4
myrun <- function() {
    cat("y:", y, "  envjn$y: ")
    print(envjn$y)
    envjn$y <- 9876
    return(0)
}
mystart()
myrun()

## y: 4   envjn$y: [1] 1
## [1] 0

cat("envjn$y:", envjn$y, "\n")

## envjn$y: 9876
```

A minor nuisance is that the special assignment <<- puts the environment object envjn into the user's workspace, rather than keeping it under the scope of mystart. One also needs to choose a name for the object that will not clash with a name likely to be used by another package. While I am sure that there may be more elegant mechanisms, the approach here has the merit of being reasonably clear, and the information stored in envjn can be used relatively safely. Thanks to Gabor Grothendieck for helpful discussions about this.

3.6 Masked (temporarily fixed) optimization parameters

Sometimes we wish to fix some of the **x** parameters. Fixing a parameter implies that there is one fewer element in the parameter vector, but adjusting all the indices is a large task, essentially a rewrite of our objective function and constraint functions. Unfortunately, although the mathematics and even the programming of the use of such fixed or **masked** parameters is relatively straightforward, it is surprisingly uncommon in optimization codes. Because it can be quite useful, I have included a chapter on masks and their use (Chapter 12) and mention some approaches to specifying fixed parameters there that are different from the following discussion.

The idea of masks has been around quite a long time. The author recalls first recognizing explicit program code in some work of Ron Duggleby of the University of Queensland, and the ideas were incorporated in Nash and Walker-Smith (1987). Masks are a special form of equality constraint. As mentioned, they present an issue of detail and tedium for nonlinear optimization programs rather than a serious mathematical obstacle.

Following Nash and Walker-Smith (1987), in my own R codes, I use an indicator vector bdmsk that puts a 0 in the position of each masked (fixed) parameter.

Initially, the other parameters are free, which is indicated by a value of 1 in the corresponding element of bdmsk. This indicator vector is also used for bounds constraints, and on return from one of the optimizers with bounds and masks capability, a bdmsk element of −3 implies a parameter is on an active lower bound, while −1 implies a parameter is on an active upper bound.

The 0's in bdmsk can be applied in a simple way to the gradient, because a fixed parameter will have zero gradient component – it is, after all, fixed. Likewise, the row and column of the Hessian corresponding to a masked parameter will be zero. While it is clear such zeros should not be an essential obstacle to an optimization method, a particular implementation of a method may inadvertently happen to create a "0/0" situation. To render an optimizer capable of handling masks is actually quite a bit simpler than introducing bounds (box constraints), but the details must nevertheless be handled carefully.

3.7   Dealing with inadmissible results

Sometimes users provide functions to define a problem, namely, the objective function, constraint functions, and sometimes derivative calculations, that gives returned values that cannot be handled by our optimization codes. The returned objects may be infinite, NULL, or NA, causing failure of the optimizer and other computations. Sometimes computations return an object that is a vector, matrix, or array when it should be numeric. For example, the sum of squares of a vector of residuals, resids, in a nonlinear least squares calculation can be efficiently found using

```
sumsqrs<-crossprod(resids)
```

Unfortunately, sumsqrs is a 1 × 1 matrix. We need to use

```
sumsqrs<-as.numeric(crossprod(resids))
```

Often such seemingly trivial matters result in a program suddenly stopping. We would ideally wish to have our program simply warn of the issue and proceed sensibly. At the very least, we would like it to terminate in a controlled manner. While it is essentially the inputs to the functions that are inadmissible and cause our troubles, it is the results of attempting a calculation in a particular context that are the source of our difficulty.

I have found that unacceptable returned values from user-written functions are one of the most common sources of failure for users in applying optimization tools to their problems. Indeed, including incorrectly computed derivatives (mostly gradients or Jacobian matrices) and poor scaling of parameters in this category explains almost all the "failures" of optimization codes I see.

Quite often users supply objective or residual functions where there are log() or sqrt() components with arguments that can become negative. The starting parameters may be such that the function is initially computable and the user may not realize that the optimization process may generate a negative argument to a function.

By writing the problem function carefully, we can indicate that inputs are inadmissible (or even undesirable). A simple expedient that often "works" is to return a very large function value. In the experimental package **optplus**, I used `badval=(0.5)*.Machine$double.xmax`, which has a value nearly 1E+308. Note, however, that simply making the function large at undesirable parameter points may distort the behavior of the optimization method. We do not want to actually use this result as a function value; it is an indicator of a point that should not be part of the domain of our function. Indeed, because it is quite possible that methods that generate points in the parameter space and use them to search in favorable directions will calculate on the returned, and very large, values, I have also used smaller settings, for example, `badval=(1e-6)*.Machine$double.xmax` which for all but very extreme functions will serve as an indicator.

In Nash and Walker-Smith (1987), we used a "noncomputability flag" that allowed optimization methods to adjust their search for an optimum. However, we wrote all our own optimizers and could react to the flag appropriately. This is not nearly so easy with optimizers written by others that operate from a different base and perspective. Very few optimization methods in R, including my own, have much capability to exploit this as yet. However, even a minimal capability allows a report of failure to be passed back to a calling program, thereby avoiding an unexpected halt in computations.

R also has the function `try()` that let us trap errors. This is used in several of my codes to avoid "crashes," but so far, it mostly does no more than report difficulties and then stop or exit. That is, we do not fully exploit the capability to backtrack from a dangerous parameter region.

3.8   Providing derivatives for functions

Many optimization tools require derivative information for the objective function. This is sufficiently important that it merits a full chapter in this book (Chapter 10). A general nonlinear objective function $f()$ will give rise to a gradient $\mathbf{g}()$. If the objective is a sum of squared residuals, then it is generally preferable to work with the residuals and find their derivatives as the Jacobian matrix and build the gradient as the inner product of the Jacobian and the vector of residuals.

Second derivative information for a general nonlinear objective function is found as the Hessian matrix $\mathbf{H}$. For a sum of squares, we find the Hessian elements as a two-term expression. Thus the j, k element of $\mathbf{H}$ is the inner product of the jth and kth columns of the Jacobian (i.e., the j, k element of the matrix $\mathbf{J}' \cdot \mathbf{J}$) plus the sum of elements of the form

$$\frac{\partial^2 r_i}{\partial x_j \partial x_k} \cdot r_i \tag{3.3}$$

Requiring the second derivatives of the residuals from users is rare, and rarer than it ought to be in the author's view. However, Hessian matrix information is commonly used in deciding if a solution is acceptable. Moreover, Hessians ought to be more commonly used in nonlinear least squares computations instead of "standard" approximations to the Jacobian of the residuals, which leave out some of the Hessian information.

Computing adequately precise derivatives is important for the proper working of some optimization tools, particularly those that make explicit use of gradient information. This is especially true of methods from the conjugate gradient /truncated Newton and quasi-Newton / variable metric families of methods, of which there are a great many algorithms and implementations, for example, Nielsen (2000), Nash (2011a), Nash (2011b), Nielsen and Mortensen (2012), Byrd et al. (1995), Dai and Yuan (1999).

For such optimization methods, it is generally considered that derivatives calculated from analytic functions are likely to work best. Such functions require either human or computer symbolic mathematics, or else automatic differentiation (there are quite good explanations at http://en.wikipedia.org/wiki/Automatic_ differentiation and http://www.autodiff.org/), which programmatically applies the chain rule and other algorithms to the program code of our function. A number of practitioners use automatic differentiation in optimization, for example, the NEOS web-based optimization system (http://www.neos-server.org/) and the AD Model Builder software (http://admb-project.org/). These have justifiably attracted a certain following, but my experience is that they require considerable effort to use. On the other hand, many workers, including some very prominent developers of optimization software, prefer to work with numerically approximated derivatives or with methods that obtain this information implicitly.

When symbolic or automatic derivatives are not available, numerical approximations are generally employed. As a rather poor example, simple forward derivative approximations can be computed by code such as

```
gr <- function(par, ...) {
    fbase <- myfn(par, ...) # ensure we have right value, may not be necessary
    df <- rep(NA, length(par))
    teps <- eps * (abs(par) + eps)
    for (i in 1:length(par)) {
        dx <- par
        dx[i] <- dx[i] + teps[i] # Dangerous step if a constraint
 is in the way!
        tdf <- (myfn(dx, ...) - fbase)/teps[i]
        if (!is.finite(tdf) || is.nan(tdf))
            tdf <- 0 # Is this a good choice?
        df[i] <- tdf
    }
    df
}
```

3.9   Derivative approximations when there are constraints

Whatever approach to derivatives is chosen, the situation is seriously magnified when there are bounds or other constraints on the parameters. In principle, we can simply project gradients onto a surface tangential to the constraint. In practice, our methods must discover if they are on or very close to a constraint surface. For my own sanity, I will not attempt to define "very close" in this context.

Worse, if we are using numerical approximations to derivatives, the computations can step out of bounds. Consider the expression for the derivative of the univariate function $f(x)$ used above in the forward approximation

$$f'(x) = \frac{f(x+h) - f(x)}{h} \tag{3.4}$$

The issue for software development is that x can be in bounds when $x + h$ is not, so we need to include checks for constraint violation as well as special code for an alternate computation of the derivative approximation. I have not seen programs that do this. The concept is fairly simple, but the actual code would be very long and detailed.

The step noted as dangerous in the above code chunk presents a challenge to programmers primarily because it imposes quite severe overheads. First, we must check *whether* the step taken for the *i*th element of the gradient crosses a bound, and if it does, we must then work out how to best approximate the gradient. For example, the code above uses a *forward* approximation because we add the axial step to the parameter vector. But equally simple are backward approximations based on subtracting the step and then using

```
tdf <- (fbase - myfn(dx, ...))/teps[i].
```

3.10   Scaling of parameters and function

Scaling of parameters for user-written functions, and occasionally, the scale of the computed function, are another issue that afflicts optimization computations. When users are aware of scaling, they are generally able to provide appropriate adjustments of parameters and functions. Automatic scaling is not trivial for nonlinear optimization because an essential feature of nonlinearity is that there is different scaling in different regions of the parameter space. Once again, this merits its own treatment in Chapter 17.

3.11   Normal ending of computations

How do we know if our computations are satisfactory? Answering this leads us to the thorny issues of **optimality tests** and **program performance**. Almost every piece of code to perform function minimization has its own particular tests for **termination** – often misidentified as **convergence**. As we have noted,

Programs terminate, while algorithms (may?) converge.

Optimality tests almost always have some relation to the well-known KKT conditions (Gill et al. 1981), but there are many approximations and shortcuts used, so it is useful to be able to actually compute the final gradient and the final Hessian. We want the gradient to be "small" and the Hessian to be "positive definite," but the meaning of these concepts in the context of real problems in floating point arithmetic are far from cut and dried. There is more on this subject in Chapter 19.

3.12   Termination tests – abnormal ending

It is very very common that optimization programs do not exit because a satisfactory answer has been found. Instead, they often return control to a calling program because too much effort has been expended, as measured by time, function evaluations, gradient evaluations, or "iterations." The last measure is unfortunate in that it is unusual to find clear, consistent documentation of what constitutes an "iteration" for a particular method.

Beyond reasons of too much effort, optimization programs may stop for a variety of reasons that are particular to the methods themselves. Let us consider one example.

R's `optim()` function includes a Nelder–Mead polytope minimizer based on the version in my book (Nash, 1979). This method, as mentioned in Section 2.10.3, takes a preliminary polytope (or simplex) made up of $n + 1$ points in n-space and modifies it according to various rules. One of these rules is called SHRINK, and it shifts n of the points toward the point with the lowest function value.

One would assume that a volume measure of the polytope would be reduced in this operation, but when I was developing the code (in BASIC on a machine with between 4K and 8K of memory for program and data, and very unsatisfactory arithmetic), I noticed that sometimes the polytope volume measure would not decrease because of rounding errors. This was generally when the polytope had collapsed onto a point that was quite often a good approximation to the minimum. Because of the possibility that the SHRINK operation did not achieve its purpose, I calculated size measures before and after the operation, making failure to reduce the polytope volume a reason to exit, as otherwise the program would cycle continuously, trying and failing to shrink the polytope.

3.13   Output to monitor progress of calculations

For large calculations, it is helpful to be able to monitor the progress of the calculations. I like to periodically see the best function value found so far and some measures of function or gradient calculations. The volume of such output is controlled by setting an appropriate option. Unfortunately, there are many names and types for such options. `trace` is a common choice of name, but it can be either a logical or an integer quantity, depending on the optimizer. There are many other names. If you are using just one method, this is hardly worth thinking about, but if you want to try different methods, the syntax changes are a large nuisance.

The volume and selection of quantities output vary greatly across optimization codes. This lack of consistency is partly due to the internals of the optimizers. However, it makes comparisons between methods much more tedious than it might be if the progress information were consistent across many optimizers. This is one of the major motivations of `optimx`.

3.14   Output of the optimization results

When optimization programs have finished, even via a successful and normal termination, the returned information may vary considerably across methods. Standardization of output was another goal of the `optimx` wrapper, and we have put in quite a lot of work to provide useful and consistent output, but there are still some awkward structures. A lot of work has been put in (thanks especially to Gabor Grothendieck) to provide tools to access the returned results from `optimx`. Similar issues exist for most large R packages that wrap several related tools.

Note that in many cases, we only want the optimal parameters. However, it is almost always worthwhile doing some examination – either automated or manual – of the function value as well as some indicator of the likely acceptability of the solution such as the KKT tests.

In rarer instances, it is important to provide the **trajectory** of the optimization. This is the set of points, possibly augmented by the function value, which mark iterations or other recognized steps in our methods. This is quite distinct from the needs of different applications for particular information to be returned or saved, which might, as we have suggested, be only the parameters and perhaps the function value and a completion code. By contrast, the trajectory might lead to an unsatisfactory "answer." What we want to know is how, and eventually why, the method got to this answer.

3.15   Controls for the optimizer

Most packages and functions include a "control" list, often called `control`. In R this list may be used to provide tolerances for convergence tests, indicators of the level of output to be provided, algorithms for computing derivatives, and similar options. Users can generally comprehend the general purpose of these controls, even if they lack the understanding to use them to best advantage.

Much more difficult for nonspecialists are controls that alter the behavior of the optimizer methods. A relatively comprehensible example of this is the pair of controls used to increase or decrease the so-called Marquardt parameter in the Levenberg–Marquardt approach to nonlinear least squares. *lambda* safeguards the Gauss–Newton search direction calculation by essentially mixing a steepest-descents component into the Gauss–Newton search direction. However, if it is too large, there is usually a slowing of progress toward a minimum. The tactic to try to work around this possibility is to decrease the Marquardt parameter *lambda*, for example, multiply it by 0.4, when we improve (that is, decrease) the sum of squares function, and increase it, for example, by a factor of 10, when we fail to get a lower sum of squares. Setting particular values for the decrease and increase controls can be important for performance, On the other hand, we want choices for these numbers, as well as the initial value of *lambda*, that are widely successful for many objective functions and starting parameters.

3.16   Default control settings

R has a good mechanism for providing default values of function arguments. This allows users to omit mention of these arguments if they are happy to use the default values provided. Thus, if we declare

```
myfn<-function(parameters, siglevel=0.05){
```

then we can use the calling syntax

```
result<-myfn(somepars)
```

and `siglevel` will be given the default value of 0.05. If we wish to override this to use a value 0.01, we put

```
result<-myfn(somepars, siglevel=0.01)
```

For the `control` list argument of a function call, we must specify the items to be particularized. Thus, for example, we might write

```
result<-myfn(somepars, control=list(trace=2))
```

to set the `trace` control, which often has a zero default.

3.17   Measuring performance

Performance is often measured by computing time, but other measures of effort are possible, such as the number of evaluations of functions, gradients, and Hessians, or "iterations" of a method. The amount of memory or number of double-precision numbers that must be stored may also be important. Unfortunately, all these are subject to the vagaries of implementation choices.

3.18   The optimization interface

Finally, there is a huge issue related to the overall way in which human users provide input to and receive output from the optimization tools. Traditionally, R is operated by typing in program lines that execute a function and return the output as a result object. Sometimes the functions we run also write data to a file or the screen. Because some early terminals were teleprinters, the command `print` is used even for writing to a screen.

As we store programs as files, we can also run them outside the R command terminal. There are tools such as `Rscript` and the construct R CMD to allow this.

Today many users are uncomfortable with having to type commands and prefer to use a pointing device such as a mouse to drag icons around a screen, choose from menu items, and generally do most of their work without touching a keyboard. R has a number of tools that allow such graphical user interfaces (GUIs) to be constructed. There are many variations on the GUI theme, each with its advantages and disadvantages for particular uses. I mentored Yixuan Qiu in a 2011 Google Summer of Code project called **optimgui** which demonstrated capabilities in this area. In particular, the interface allowed the objective function to be coded (we typed code into a template), data to be provided by selection, and the function executed for a set of parameters. Testing and checking were actions we wanted to make default behavior. The user could easily select an optimizer and attempt to run it to solve the problem.

As with most optimization tools, this was relatively easy to get running to a demonstration level. The unseen bulk of the iceberg is, however, the many small but critical checks for things that are not or should not be allowed, things that will make the results unsatisfactory. Thus **optimgui** is "working" but unfinished at the time of writing and is not part of the CRAN repository.

References

Byrd RH, Lu P, Nocedal J and Zhu CY 1995 A limited memory algorithm for bound constrained optimization. SIAM Journal on Scientific Computing 16(5), 1190–1208.

Dai YH and Yuan Y 1999 A nonlinear conjugate gradient method with a strong global convergence property. SIAM Journal on Optimization 10, 177–182.

Feldman SI, Gay DM, Maimone MW and Schryer NL 1991 Availability of f2c - a fortran to c converter. SIGPLAN Fortran Forum 10(2), 14–15.

Gill PE, Murray W and Wright MH 1981 Practical Optimization. Academic Press, London.

Murphy L 2012 likelihood: Methods for maximum likelihood estimation. R package version 1.5.

Nash JC 1979 Compact Numerical Methods for Computers: Linear Algebra and Function Minimisation. Adam Hilger, Bristol. Second Edition, 1990, Bristol: Institute of Physics Publications.

Nash JC 2011a **Rcgmin**: Conjugate Gradient Minimization of Nonlinear Functions with Box Constraints. Nash Information Services Inc. R package version 2011-2.10.

Nash JC 2011b **Rvmmin**: Variable Metric Nonlinear Function Minimization with Bounds Constraints. Nash Information Services Inc. R package version 2011-2.25.

Nash JC and Walker-Smith M 1987 Nonlinear Parameter Estimation: An Integrated System in BASIC. Marcel Dekker, New York. See http://www.nashinfo.com/nlpe.htm for an expanded downloadable version.

Nielsen HB 2000 UCMINF - an algorithm for unconstrained, nonlinear optimization. Technical report, Department of Mathematical Modelling, Technical University of Denmark. Report IMM-REP-2000-18.

Nielsen HB and Mortensen SB 2012 ucminf: General-purpose unconstrained non-linear optimization. R package version 1.1-3.

4

One-parameter root-finding problems

Quite often problems involve only one parameter. In such cases, some general computational tools actually "crash" when asked to solve them, although this is generally a failure in their software design to deal with the single dimension. However, even if we have a well-programmed solver for any number of parameters, it is usually a good idea to use a tool for finding the root of a function of one parameter or to find the minimum (or maximum) of such a function rather than try to apply the more general program. This chapter considers root-finding.

4.1 Roots

Let us look at how problems involving the root(s) of a function of one variable may arise and how R may solve them. Although we will mention polynomial root-finding, because it is a very common problem, we regard this particular problem (and eigenvalues of matrices) to be somewhat different from those that will be the focus here.

We also wish to point out the limitations of computational technology for root-finding. Treating root-finders as black boxes is, in the author's view, dangerous, in that it risks many possibilities for poor approximations to the answers we desire, or even drastically wrong answers. Largely, this is because users may make assumptions about the problems and/or the software that are not justified. Indeed, in the present treatment, we mostly seek only real-valued solutions to equations – a big assumption.

We also want to show that the built-in tool for one-dimensional root-finding in R (uniroot()), while a very good choice, is still based on several design considerations that may not be a good fit to particular user problems.

Nonlinear Parameter Optimization Using R Tools, First Edition. John C. Nash.
© 2014 John Wiley & Sons, Ltd. Published 2014 by John Wiley & Sons, Ltd.
Companion Website: www.wiley.com/go/nonlinear_parameter

4.2   Equations in one variable

There are many mathematical problems stated as equations. If we have one equation and only one variable in the equation is "unknown," that is, not yet determined, then we should be able to solve the equation. In order to do this, we shall specify that the unknown variable is x and rewrite the equation as

$$f(x) = 0 \tag{4.1}$$

where $f()$ is some function of the single parameter.

Of course, there may be more than one value of x that causes $f(x)$ to be zero. This multiplicity of solutions is one of the principal difficulties for root-finding software. Roots might also have complex values, and it is quite reasonable that users (and software) may consider that only real roots are admissible and wanted. As mentioned, we will seek real roots unless we state otherwise.

4.3   Some examples

Examples are often the most straightforward way to learn about a subject.

4.3.1   Exponentially speaking

The exponential function $\exp(-\text{alpha} * x)$ descends to an asypmtote at 0 with a rate dependent on the positive parameter alpha. Thus the function

$$efn(x) = \exp(-\text{alpha} * x) - 0.2 \tag{4.2}$$

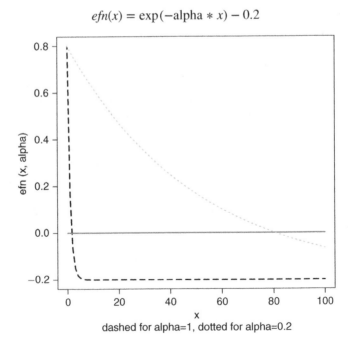

dashed for alpha=1, dotted for alpha=0.2

Figure 4.1

has a root at

$$x^* = -\log(0.2)/\text{alpha} \tag{4.3}$$

For small alpha, the root will be large, as the function will be very "flat" as it crosses zero. For alpha moderate to large, the function will be very rapidly changing as it crosses the y-axis near $x = 0$. This problem provides a simple example of some of root-finding and a quite good test of software. Of course, we already have the answer and can see it clearly if we draw the *efn* function and also draw a line for x-axis. Let us look at the cases alpha = 1 and alpha = 0.02. The *efn* function is drawn in dashed and dotted, respectively, by the code that follows. Note that an attempt to find a root in an interval where the end points do not have function values with different signs gives an error (Figure 4.1).

```
cat("exponential example\n")

## exponential example

require("rootoned")
alpha <- 1
efn <- function(x, alpha) {
    exp(-alpha * x) - 0.2
}
zfn <- function(x) {
    x * 0
}
tint <- c(0, 100)
curve(efn(x, alpha = 1), from = tint[1], to = tint[2], lty = 2,
ylab = "efn(x, alpha)")
title(sub = "Dashed for alpha = 1, dotted for alpha = 0.2")
curve(zfn, add = TRUE)
curve(efn(x, alpha = 0.02), add = TRUE, lty = 3)

rform1 <- -log(0.2)/1
rform2 <- -log(0.2)/0.02
resr1 <- uniroot(efn, tint, tol = 1e-10, alpha = 1)
cat("alpha = ", 1, "\n")

## alpha =  1

cat("root(formula)=", rform1, "  root1d:", resr1$root, " tol=",
resr1$estim.prec,
    " fval=", resr1$f.root, "  in   ", resr1$iter, "\n\n")

## root(formula)= 1.609    root1d: 1.609  tol= 0.0005702    fval= 0    in    13

resr2 <- uniroot(efn, tint, tol = 1e-10, alpha = 0.02)
cat("alpha = ", 0.02, "\n")

## alpha =  0.02

cat("root(formula)=", rform2, "  root1d:", resr2$root,
" tol=", resr2$estim.prec,
    " fval=", resr2$f.root, "  in   ", resr2$iter, "\n\n")

## root(formula)= 80.47   root1d: 80.47 tol= 5.004e-11  fval= 9.381e-15  in  8

cat("\n")
```

```
cat("Now look for the root in [0,1]\n")

## Now look for the root in [0,1]

tint2a <- c(0, 1)
resr2a <- uniroot(efn, tint2a, tol = 1e-10, alpha = 0.02)

## Error: f() values at end points not of opposite sign
```

4.3.2 A normal concern

The problem we will now consider is actually one of finding the maximum of a one-dimensional problem, the subject of a related chapter. Its relevance to the root-finding problem is how it illustrates the difficulties that numeric precision can introduce.

Consider the Gaussian (normal) density function

$$f(x) = \frac{1}{\text{sqrt}(2 * \text{pi} * \text{sigma}^2)} \exp(-0.5 * ((x - \text{mu})/\text{sigma})^2) \qquad (4.4)$$

This is the usual "bell-shaped" curve. It is always positive, and we want to find its maximum. Obviously, we know the answer – the mean mu. As we are discussing root-finding, we will look at the derivative with respect to x. This function can be found using R with the D() or deriv() tools, which we will explore in Chapter 5.

The function we want to use is

$$g(x) = \frac{-1}{\text{sqrt}(2 * \text{pi} * \text{sigma}^2))} * \frac{(x - \text{mu})}{\text{sigma}^2} \exp(-(x - \text{mu})^2/(2 * \text{sigma}^2)) \quad (4.5)$$

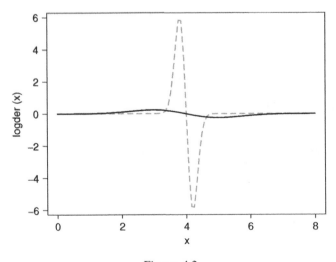

Figure 4.2

Let us set mu = 4 and draw the function from 0 to 8 for sigma = 1 (in gray) and sigma = 0.1 (Figure 4.2). To keep the graph viewable, we the log of the magnitudes but keep the sign. As the standard deviation sigma gets smaller, the function gets steeper.

Let us use 2 and 6 as the limits of our search interval and find the root for *sigma* progressively smaller.

```
cat("Gaussian derivative\n")

## Gaussian derivative

der <- function(x, mu, sigma) {
    dd <- (-1) * (1/sqrt(2 * pi * sigma^2))*(exp(-(x - mu)^2/(2 * sigma^2)) *
        ((x - mu)/(sigma^2)))
}
r1 <- uniroot(der, lower = 1, upper = 6, mu = 4, sigma = 1)
r.1 <- uniroot(der, lower = 1, upper = 6, mu = 4, sigma = 0.1)
r.01 <- uniroot(der, lower = 1, upper = 6, mu = 4, sigma = 0.01)
r.001 <- uniroot(der, lower = 1, upper = 6, mu = 4, sigma = 0.001)
sig <- c(1, 0.1, 0.01, 0.001)
roo <- c(r1$root, r.1$root, r.01$root, r.001$root)
tabl <- data.frame(sig, roo)
print(tabl)

##      sig roo
## 1 1.000   4
## 2 0.100   4
## 3 0.010   1
## 4 0.001   1
```

What is going on here?! We know the root should be at 4, but in the final two cases, it is at 1. The reason turns out to be that the function value at $x = 1$ is computationally zero – a root. The danger is that we just do not recognize that this is not the "root" we likely wish to find, for example, if we are using the derivative as a way to sharpen the finding of a peak on a spectrometer.

Fortunately, thanks to Martin Maechler, there is the package **Rmpfr** (Maechler, 2013) that allows us to use extended precision arithmetic. This will also work with many R functions, but not with tools such as uniroot() that are programmed in languages other than R. However, when I learned of this issue at the 2011 UseR! meeting, I spent some time in one of the sessions that did not overlap my interests in coding an R version of the root-finder (zeroin() in the developmental R-forge package **rootoned** from project **optimizer**). Martin included a modified version of this as unirootR() in **Rmpfr**. This shows quite clearly that there is no root at 1.

Unfortunately, the code to present this example could not be run in the **knitr** environment that generated much of this book. (We are still trying to discover the reason.) Thus the computations were carried out in the R terminal and the results are copied here.

```
> require(Rmpfr)
> der<-function(x,mu,sigma){
+      dd<-(-1)*(1/sqrt(2 * pi * sigma^2)) * (exp(-(x - mu)^2/(2 * sigma^2)) *
+      ((x - mu)/(sigma^2)))
+ }
> rootint1<-mpfr(c(1.9, 2.1),200)
> rootint1
2 'mpfr' numbers of precision  200   bits
[1] 1.8999999999999999118215802998747676766109466552734375
[2] 2.1000000000000000888178419700125232333890533447265625

> r140<-try(unirootR(der, rootint1, tol=1e-40, mu=4, sigma=1))
> # r140

> rootint2<-mpfr(c(1, 6),200)
> rootint2
2 'mpfr' numbers of precision  200   bits
[1] 1 6

> r140a<-try(unirootR(der, rootint2, tol=1e-40, mu=4, sigma=1))
> r140a
$root
1 'mpfr' number of precision  200   bits
[1] 3.9999999999999999999999999999999999999999999999994225054110699770

$f.root
1 'mpfr' number of precision  200   bits
[1] 2.3038700822723117720200096647152975711858994581054996741312416e-47

$iter
[1] 8

$estim.prec
[1] 5e-41

$converged
[1] TRUE
```

4.3.3 Little Polly Nomial

Polynomial roots are a common problem that should generally be solved by methods different from those where our goal is to find a single real root of a real scalar function of one variable. This is easily illustrated by an example. Suppose we want the roots of the polynomial

$$z(x) = 10 - 3 * x + x^3 \qquad (4.6)$$

This has the set of coefficients that we would establish using the R code

```
    z <- c(10, -3, 0, 1)
```
The package **polynom** solves this easily using polyroot(z).

```
z <- c(10, -3, 0, 1)
simpol <- function(x) {
    # calculate polynomial z at x
    zz <- c(10, -3, 0, 1)
    ndeg <- length(zz) - 1 # degree of polynomial
    val <- zz[ndeg + 1]
    for (i in 1:ndeg) {
        val <- val * x + zz[ndeg + 1 - i]
    }
    val
}
tint <- c(-5, 5)
cat("roots of polynomial specified by ")

## roots of polynomial specified by

print(z)

## [1] 10 -3  0  1

require(polynom)
allroots <- polyroot(z)
print(allroots)

## [1]   1.3064+1.4562i  -2.6129+0.0000i   1.3064-1.4562i

cat("single root from uniroot on interval ", tint[1], ",", tint[2], "\n")

## single root from uniroot on interval  -5 , 5

rt1 <- uniroot(simpol, tint)
print(rt1)

## $root
## [1] -2.6129
##
## $f.root
## [1] 2.5321e-06
##
## $iter
## [1] 9
##
## $estim.prec
## [1] 6.1035e-05
```

Here we see that uniroot does a perfectly good job of finding the real root of this cubic polynomial. It will not, however, solve the simple polynomial

$$y(x) = (x - 4)^2 \qquad (4.7)$$

because both the roots are at 4 and the function never crosses the y-axis. uniroot() insists on having a starting interval where the function has opposite sign at the ends of the interval. `uniroot()` should (and in our example below does) work when there is a root with odd multiplicity, so that there is a crossing of the axis.

```
z <- c(16, -8, 1)
simpol2 <- function(x) {
    # calculate polynomial z at x
    val <- (x - 4)^2
}
tint <- c(-5, 5)
cat("roots of polynomial specified by ")

## roots of polynomial specified by

print(z)

## [1] 16 -8  1

require(polynom)
allroots <- polyroot(z)
print(allroots)

## [1] 4+0i 4-0i

cat("single root from uniroot on interval ", tint[1], ",", tint[2], "\n")

## single root from uniroot on interval  -5 , 5

rt1 <- try(uniroot(simpol2, tint), silent = TRUE)
print(strwrap(rt1))

## [1] "Error in uniroot(simpol2, tint) : f() values at end points not of"
## [2] "opposite sign"

cat("\n Try a cubic\n")

##
##   Try a cubic

cub <- function(z) {
    val <- (z - 4)^3
}
cc <- c(-64, 48, -12, 1)
croot <- polyroot(cc)
croot

## [1] 4-0i 4+0i 4+0i

ans <- uniroot(cub, lower = 2, upper = 6)
ans

## $root
## [1] 4
##
## $f.root
## [1] 0
##
```

```
## $iter
## [1] 1
##
## $estim.prec
## [1] 2
```

4.3.4   A hypothequial question

In Canada, mortgages fall under the Canada Interest Act. This has an interesting clause:

> 6. Whenever any principal money or interest secured by mortgage of real estate is, by the mortgage, made payable on the sinking fund plan, or on any plan under which the payments of principal money and interest are blended, or on any plan that involves an allowance of interest on stipulated repayments, no interest whatsoever shall be chargeable, payable, or recoverable, on any part of the principal money advanced, unless the mortgage contains a statement showing the amount of such principal money and the rate of interest thereon, calculated yearly or half-yearly, not in advance.

This clause allowed the author to charge rather a high fee to recalculate the payment schedule for a private lender who had, in a period when the annual rate of interest was around 20% in the early 1980s, used a US computer program. The borrower wanted to make monthly payments. This is feasible in Canada, but we must do the calculations with a rate that is equivalent to the half-yearly rate. So if the annual rate is $R\%$, the semiannual rate is $R/2\%$, and we want compounding at a monthly rate that is equal to this semiannual rate. That is,

$$(1 + I/100)^6 = (1 + R/200) \tag{4.8}$$

and our root-finding problem is the solution of

$$A(I) = (1 + I/100)^6 - (1 + R/200) = 0 \tag{4.9}$$

We can actually write down a solution, of course, as

$$I = 100 * ((1 + R/200)^{(1/6)} - 1) \tag{4.10}$$

A textbook (and possibly dangerous) approach to this has been to plug this formula into a spreadsheet or other system (including R). The difficulty is that approximations to the fractional (1/6) power are just that, approximations. And the answer for low interest rates are bound to give digit cancellation when we subtract 1 from the (1/6) root of a number not very different from 1. However, R does, in fact, a good job. Using `uniroot` on the function $A(I)$ above is also acceptable if the tolerance is specified and small, otherwise we get a rather poor answer.

A quite good way to solve this problem is by means of the binomial theorem, where the expansion of $(1 + h)^{(1/6)}$ (substituting h for $R/200$) is

$$(1 + h)^{(1/6)} = 1 + (1/6) * h + (1/6)((1/6) - 1) * h^2/2! +$$
$$(1/6)((1/6) - 1)((1/6) - 2) * h^3/3! + \cdots \qquad (4.11)$$

Besides converging very rapidly, this series expansion begins with 1, which we can subtract analytically, thereby avoiding digit cancellation. The following calculations present the ideas in program form, which show that any of the three approaches are reasonable for this problem.

```
mrate <- function(R) {
    val <- 0
    den <- 1
    fact <- 1/6
    term <- 1
    rr <- R/200
    repeat {
        # main loop
        term <- term * fact * rr/den
        vallast <- val
        val <- val + term
        # cat('term =',term,' val now ',val,'\n')
        if (val == vallast)
            break
        fact <- (fact - 1)
        den <- den + 1
        if (den > 1000)
            stop("Too many terms in mrate")
    }
    val * 100
}
A <- function(I, Rval) {
    A <- (1 + I/100)^6 - (1 + R/200)
}

Rvec <- c()
i.formula <- c()
i.root <- c()
i.mrate <- c()

for (r2 in 0:18) {
    R <- r2/2 # rate per year
    Rvec <- c(Rvec, R)
    i.formula <- c(i.formula, 100 * ((1 + R/200)^(1/6) - 1))
    i.root <- c(i.root, uniroot(A, c(0, 20), tol = .Machine$double.eps,
Rval = R)$root)
    i.mrate <- c(i.mrate, mrate(R))
}
tabR <- data.frame(Rvec, i.mrate, (i.formula - i.mrate), (i.root - i.mrate))
colnames(tabR) <- c("rate", "i.mrate", "form - mrate", "root - mrate")
print(tabR)

##    rate i.mrate form - mrate root - mrate
## 1   0.0 0.000000   0.0000e+00   0.0000e+00
## 2   0.5 0.041623  -3.1572e-15   7.8132e-15
```

```
## 3    1.0 0.083160    4.7323e-15   -6.3144e-15
## 4    1.5 0.124611   -5.1625e-15    5.7038e-15
## 5    2.0 0.165976    5.5511e-16    3.3307e-16
## 6    2.5 0.207256    9.9920e-16    2.6923e-15
## 7    3.0 0.248452    5.1625e-16   -5.8287e-15
## 8    3.5 0.289562    1.6653e-16   -2.1094e-15
## 9    4.0 0.330589    3.7192e-15   -7.2164e-15
## 10   4.5 0.371532   -6.8279e-15    4.1078e-15
## 11   5.0 0.412392   -3.7192e-15    7.1054e-15
## 12   5.5 0.453168    7.6605e-15   -3.2196e-15
## 13   6.0 0.493862   -9.5479e-15    1.1657e-15
## 14   6.5 0.534474    5.5511e-15   -5.3291e-15
## 15   7.0 0.575004   -1.3323e-15   -5.8842e-15
## 16   7.5 0.615452   -5.3291e-15    5.3291e-15
## 17   8.0 0.655820    6.8834e-15   -3.9968e-15
## 18   8.5 0.696106   -8.5487e-15    2.2204e-15
## 19   9.0 0.736312    1.1102e-15   -9.5479e-15
```

4.4   Approaches to solving 1D root-finding problems

From the above examples, it is clear that any root-finder needs to tell the users what it is intended to do. Unfortunately, few do make clear their understandings of a problem. And users often do not read such information anyway! Let us consider the main ways in which root-finders could be called.

- We could require the user to supply an interval, a lower and an upper bounds for a prospective root, and further insist that the function values at the ends of the interval have opposite sign. This is the situation for R's uniroot() (Brent, 1973) and root1d() (Nash, 1979) from the package **rootoned** at https://r-forge.r-project.org/R/?group_id=395.

- An alternative approach is to consider that we will search for a root from a single value of the argument x of our function $f(x)$. This is the setup for a one-dimensional Newton's method, which iterates using the formula

$$x_{\text{new}} = x - f(x)/f'(x) \tag{4.12}$$

- We may wish to specify the function with no starting value of the argument x. An example is the polyroot function that we have already used. As far as the author is aware, there are no general root-finders of this type in R, or elsewhere to his knowledge.

With R, a possible package could be developed using an **expression** for the function. This would allow the symbolic differentiation operator D() to be applied to get the derivative, although there are some limitations to this tool. For example, it does not comprehend sum() and other fairly common functions. I have not bothered to pursue this possibility yet.

It is also possible to consider a secant-rule version of Newton's method. That is, we use two points as in zeroin(), root1d, or uniroot but do not require that they have function values of opposite sign. An alternative setup would use an initial guess to the root and a stepsize.

4.5   What can go wrong?

Root-finding may seem like a fairly straightforward activity, but there are many situations that cause trouble.

Multiple roots are common for polynomial root-finding problems. We already have seen cases in which the polynomial is a power of a monomial $(x - \text{root})$. Computationally, such problems are "easy" if we have an odd power and are using a root-finder for which we supply an interval that brackets the root. We will not, however, learn the multiplicity of the root without further work.

More troublesome are such problems when the function just touches zero, as when the power of the monomial is even. In such case, Newton's method can sometimes succeed, but there is a danger of numerical issues if $f'(x)$ becomes very small, so that the iteration formula blows up. In fact, Newton's methods generally need to be safeguarded against such computational issues. Most of the code for a successful Newton-like method will be in the safeguards; the basic method is trivial but prone to numerical failure. Note that the **Rmpfr** tools may be of help, but because they require much more computational effort, they need to be used judiciously.

Problems with no root at all can also occur, as in the following example, which computes the tangent of an angle given in degrees. We also supply gradient code.

```
mytan <- function(xdeg) {
    # tangent in degrees
    xrad <- xdeg * pi/180 # conversion to radians
    tt <- tan(xrad)
}
gmytan <- function(xdeg) {
    # tangent in degrees
    xrad <- xdeg * pi/180 # conversion to radians
    gg <- pi/(180 * cos(xrad)^2)
}
```

Let us consider some example output, which has been edited to avoid excessive space, where we seek a root between 80° and 100°. Here the function has a singularity, but its well-defined values at the initial limits of the interval provided are of opposite signs. Graphing the function, assuming we do not get an error in computing it, will show the issue clearly. However, we could see what happens with four different root-finders, uniroot() and the three root-finders from **rootoned**, namely, root1d(), zeroin(), and newt1d(). For the last routine, we include the iteration output, noting that

```
tint <- c(80, 100)
ru <- uniroot(mytan, tint)
ru
```

```
## $root
## [1] 90
##
## $f.root
## [1] -750992
##
## $iter
## [1] 19
##
## $estim.prec
## [1] 7.6293e-05

rz <- zeroin(mytan, tint)
rz

## $root
## [1] 90
##
## $froot
## [1] -6152091027
##
## $rtol
## [1] 9.3132e-09
##
## $maxit
## [1] 30

rr <- root1d(mytan, tint)

## Warning: Final function magni-
tude > 0.5 * max(abs(f(lbound)), abs(f(ubound)))

rr

## $root
## [1]100
##
## $froot
## [1] -5.6713
##
## $rtol
## [1] 0
##
## $fcount
## [1] 6

rn80 <- newt1d(mytan, gmytan, 80)
rn80

## $root
## [1] 0
##
## $froot
## [1] -5.6395e-29
##
## $itn
## [1] 8
```

```
rn100 <- newt1d(mytan, gmytan, 100)
rn100

## $root
## [1] 180
##
## $froot
## [1] -1.2246e-16
##
## $itn
## [1] 8
```

Notes:

- `zeroin()` and `uniroot()` do not give any warning of trouble and return a "root" near 90°, but with a function value at that argument having a large magnitude;

- `root1d()` does give a warning of this occurrence;

- `newt1d()`, started at an argument of 80° (one end of the starting interval) goes to 0° (another "root") that is outside the starting interval used for the other root-finders. Similarly, starting at 100° returns a root at 180°. These results are, of course, correct.

4.6   Being a smart user of root-finding programs

Mostly – and I am cowardly enough not to define "mostly" – users of root-finders such as `uniroot()` get satisfactory results without trouble. On the other hand, it really is worthwhile checking these results from time to time. This is easily done with the `curve` function that lets one draw the function of interest. Examples above show how to add a horizontal line at 0 to provide a reference and make checking the position of the root easy. Even within codes, it is useful to generate a warning if the function value at the proposed root is "large," for example, of the same order of magnitude as the function values at the ends of the initial interval for the search. Indeed, I am surprised `uniroot()` does not have such a warning and have put such a check into one of my own routines.

It is also worth remembering the package **Rmpfr** when issues of precision arise, although this recommendation comes with the caveats that code may need to be changed so that non-R functions are not called and that both the performance and the output may become unattractive.

4.7   Conclusions and extensions

From the discussion earlier,

- Methods for univariate root-finding work efficiently well but still need "watching". This applies to almost any iterative computation.

- There is a need for more "thoughtful" methods that give a user much more information about his or her function and suggest potential issues to be investigated. Such tools would be intended for use when the regular tools appear to be giving inappropriate answers.

- As always, more good test cases and examples are useful to improve our methods.

References

Brent R 1973 Algorithms for Minimization without Derivatives. Prentice-Hall, Englewood Cliffs, NJ.

Maechler M 2013 *Rmpfr: R MPFR - Multiple Precision Floating-Point Reliable*. R package version 0.5-3.

Nash JC 1979 Compact Numerical Methods for Computers: Linear Algebra and Function Minimisation. Hilger, Bristol.

5

One-parameter minimization problems

Problems that involve minimizing a function of a single parameter are less common in themselves than as subproblems within other optimization problems. This is because many optimization methods generate a search direction and then need to find the "best" step to take along that direction. There are, however, problems that require a single parameter to be optimized, and it is important that we have suitable tools to solve these.

5.1 The `optimize()` function

R has a built-in function to find the minimum of a function $f(x)$ of a single parameter x in a given interval $[a, b]$. This is `optimize()`. It requires a function or expression to be supplied, along with an interval specified as a two-element vector, as in the following example. The method is based on the work by Brent (1973).

A well-known function in molecular dynamics is the Lennard–Jones 6–12 potential (http://en.wikipedia.org/wiki/Lennard-Jones_potential). Here we have chosen a particular form as the function is extremely sensitive to its parameters ep and sig. Moreover, the limits for the curve() have been chosen so that the graph is usable because the function has extreme scale changes (Figure 5.1). Let us see if we can find the minimum.

```
# The L-J potential function
ljfn <- function(r, ep = 0.6501696, sig = 0.3165555) {
    fn <- 4 * ep * ((sig/r)^12 - (sig/r)^6)
}
min <- optimize(ljfn, interval = c(0.001, 5))
print(min)
```

Nonlinear Parameter Optimization Using R Tools, First Edition. John C. Nash.
© 2014 John Wiley & Sons, Ltd. Published 2014 by John Wiley & Sons, Ltd.
Companion Website: www.wiley.com/go/nonlinear_parameter

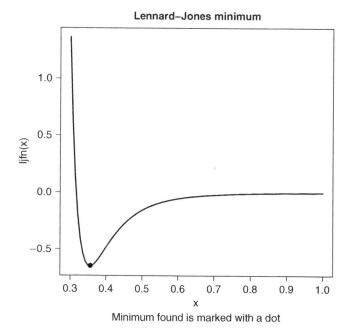

Figure 5.1

```
## $minimum
## [1] 0.35531
##
## $objective
## [1] -0.65017

## Now graph it
curve(ljfn, from = 0.3, to = 1)
points(min$minimum, min$objective, pch = 19)
title(main = "Lennard Jones minimum")
title(sub = "Minimum found is marked with a dot")
```

Later, in Chapter 11, we will see a way to use `optimize()` in a way that avoids the dangers of a zero-divide.

5.2 Using a root-finder

As we have seen in Chapter 4, if we have functional form of the derivative of `ljfn`, we can attempt to solve for the root of this function and thereby hope to find an extremum of the original function. This does not, of course, guarantee that we will find a minimum, because the gradient can be zero at a maximum or saddle point. Here we use `uniroot`, but any of the ideas from Chapter 4 can be used.

```
ljgr <- function(r, ep = 0.6501696, sig = 0.3165555) {
    expr1 <- 4 * ep
    expr2 <- sig/r
    expr8 <- sig/r^2
    value <- expr1 * (expr2^12 - expr2^6)
    grad <- -(expr1 * (12 * (expr8 * expr2^(11)) - 6 * (expr8 * expr2^5)))
}
root <- uniroot(ljgr, interval = c(0.001, 5))
cat("f(", root$root, ")=", root$f.root, " after ", root$iter, " Est. Prec'n=",
    root$estim.prec, "\n")

## f( 0.3553 )= -0.0079043   after   13   Est. Prec'n= 6.1035e-05
```

5.3   But where is the minimum?

This is all very good, but most novices to minimization soon realize that they need
to provide an `interval` to get a sensible answer. In many problems – and most
of the ones I have had to work with – this is the hard part of getting an answer.
Consider the following example (Figure 5.2).

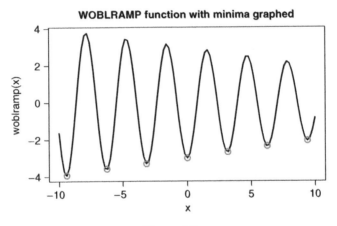

Figure 5.2

```
woblramp <- function(x) {
    (-3 + 0.1 * x) * cos(x * 2)
}
require(pracma, quietly = TRUE)
wmins <- findmins(woblramp, a = -10, b = 10)
curve(woblramp, from = -10, to = 10)
ymins <- woblramp(wmins) # compute the function at minima
points(wmins, ymins, col = "red") # and overprint on graph
title(main = "WOBLRAMP function with minima graphed")
```

An approach to this is to perform a grid search on the function. That is what Borchers (2013) does with function findmins from package **pracma**. Another approach is to use the base R function curve and simply draw the function between two values of the argument, which we did earlier. Indeed, drawing a 1D function is natural and well advised as a first step. We will then soon discover those nasty scaling issues that get in the way of automated solutions of problems of this sort.

In the earlier example, I considered issuing the require statement without the quietly=TRUE condition. This directive generally suppresses message about the replacement of function already loaded with those of the same name in **pracma**. The following is an example showing that possibility.

```
detach(package:pracma)
require(numDeriv)

## Loading required package: numDeriv

require(pracma)

## Loading required package: pracma
##
## Attaching package: 'pracma'
##
## The following objects are masked from 'package:numDeriv':
##
##     grad, hessian, jacobian
```

I believe the confusion of functions that have the same name is quite often a source of trouble for users of R. If one is in doubt, it is possible to invoke functions using the form

packagename::functionname()

and I recommend this rather cumbersome approach if there is any suspicion that there may be confusion.

5.4   Ideas for 1D minimizers

The literature of both root-finding and minimization of functions of one parameter was quite rich in the 1960s and 1970s. Essentially, developers have sought to

- provide a robust method that will find at least the minimum of a unimodal function, and hopefully one minimum of a multimodal function, when given a starting interval for the parameter;

- be fast, that is, use few function and/or derivative evaluations when the function is "well behaved;"

- be relatively easy to implement.

To this end, the methods try to meld robust searches such as success–failure search (Dixon, 1972), Golden-section search (http://en.wikipedia.org/wiki/Golden_section_search), or bisection (http://en.wikipedia.org/wiki/Bisection_method) with an inverse interpolation based on a simple polynomial model, usually a quadratic. One of the most popular is that of Brent (1973), called *FMIN*.

For most users, even those with quite complicated functions, 1D minimization is not likely to require heroic efforts to improve efficiency. When there are thousands of problems to solve, then it is perhaps worth looking at the method(s) a bit more closely. In such cases, good starting value estimates and customized termination tests may do more to speed up the computations than changing method.

On the other hand, the method requirements may be important. The Brent/Dekker codes ask us to specify an interval for the parameter of interest. A code using the success/failure search (generally followed by quadratic inverse interpolation) uses a starting point and a step for the parameter of interest. These different approaches force the user to provide rather different information. If we have a function that involves fairly straightforward computations, but no idea of the location of the minimum, then the "point + step" formulation is much less demanding in my experience, because we do not need to worry that we have bracketed a minima.

One matter that users must keep in mind is that many functions are not defined for the entire domain of the parameter. For example, consider the function

$$f(x) = \log(x) * (x^2 - 3000) \tag{5.1}$$

This function is undefined for x nonpositive. It is likely sensible for the user to add a test for this situation and return a very large value for the function. This will avoid trouble but may slow down the methods. Note that the R function optimize() does seem to have such protection built-in, but users should **not** count on such careful coding of minimizers. The following is an illustration of how to provide the protection.

```
# Protected log function
mlp <- function(x) {
    if (x < 1e-09)
        fval <- 0.1 * .Machine$double.xmax else fval <- (log(x) *
    (x * x - 3000))
}
ml <- function(x) {
    fval <- (log(x) * (x * x - 3000))
}
min <- optimize(mlp, interval = c(-5, 5))
print(min)
```

```
## $minimum
## [1] 4.9999
##
## $objective
## [1] -4788

minx <- optimize(ml, interval = c(-5, 5))

## Warning: NaNs produced
## Warning: NA/Inf replaced by maximum positive value

print(minx)

## $minimum
## [1] 4.9999
##
## $objective
## [1] -4788

## But we missed the solution!
min2 <- optimize(mlp, interval = c(-5, 55))
print(min2)

## $minimum
## [1] 20.624
##
## $objective
## [1] -7792.1
```

The failure of optimize() to tell us that the minimum has not been found is a weakness that I believe is unfortunate. We can easily check if the function has a small gradient. However, it is also suspicious that the reported minimum is at the limit of our supplied interval.

Adding better safeguards to a version of optimize() is one of the tasks on my "to do" list.

5.5   The line-search subproblem

One-dimensional minimization occurs very frequently in multiparameter optimizations, where a subproblem is to find "the minimum" of the function along a search direction from a given point. In practice, because the search direction is likely to be imperfect, a precise minimum along the line may not be required, and many line-search subprocesses seek only to provide a satisfactory improvement. Note also that the "point + step" formulation is more convenient than the interval start approach for the line search, although interval approaches may have their place in trust region methods where there is a limit to how far we want to proceed along a particular search direction.

References

Borchers HW 2013 *pracma: Practical Numerical Math Functions*. R package version 1.5.0.

Brent R 1973 Algorithms for Minimization without Derivatives. Prentice-Hall, Englewood Cliffs, NJ.

Dixon L 1972 Nonlinear Optimisation. The English Universities Press Ltd., London.

6

Nonlinear least squares

As we have mentioned, the nonlinear least squares problem is sufficiently common and important that special tools exist for its solution. Let us look at the tools R provide either in the base system or otherwise for its solution.

6.1 `nls()` from package `stats`

In the commonly distributed R system, the **stats** package includes `nls()`. This function is intended to solve nonlinear least squares problems, and it has a large repertoire of features for such problems. A particular strength is the way in which `nls()` is called to compute nonlinear least squares solutions. We can specify our nonlinear least squares problem as a mathematical expression, and `nls()` does all the work of translating this into the appropriate internal computational structures for solving the nonlinear least squares problem. In my opinion, `nls()` points the way to how nonlinear least squares and other nonlinear parameter estimation should be implemented and is a milestone in the software developments in this field. Thanks to Doug Bates and his collaborators for this.

  `nls()` does, unfortunately, have a number of shortcomings, which are discussed in the following text. We also show some alternatives that can be used to overcome the deficiencies.

6.1.1 A simple example

Let us consider a simple example where `nls()` works using the weight loss of an obese patient over time (Venables and Ripley, 1994, p. 225) (Figure 6.1). The data is in the R package **MASS** that is in the base distribution under the name `wtloss`. I have extracted that here so that it is not in a separate file from the manuscript.

Nonlinear Parameter Optimization Using R Tools, First Edition. John C. Nash.
© 2014 John Wiley & Sons, Ltd. Published 2014 by John Wiley & Sons, Ltd.
Companion Website: www.wiley.com/go/nonlinear_parameter

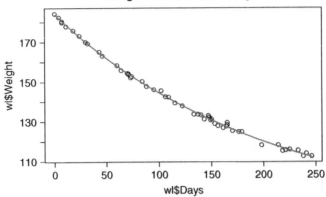

Figure 6.1

```
Weight <- c(184.35, 182.51, 180.45, 179.91, 177.91, 175.81, 173.11, 170.06, 169.31,
       165.1, 163.11, 158.3, 155.8, 154.31, 153.86, 154.2, 152.2, 152.8, 150.3, 147.8,
       146.1, 145.6, 142.5, 142.3, 139.4, 137.9, 133.7, 133.7, 133.3, 131.2, 133, 132.2,
       130.8, 131.3, 129, 127.9, 126.9, 127.7, 129.5, 128.4, 125.4, 124.9, 124.9, 118.2,
       118.2, 115.3, 115.7, 116, 115.5, 112.6, 114, 112.6)
Days <- c(0, 4, 7, 7, 11, 18, 24, 30, 32, 43, 46, 60, 64, 70, 71, 71, 73, 74, 84,
       88, 95, 102, 106, 109, 115, 122, 133, 137, 140, 143, 147, 148, 149, 150, 153,
       156, 161, 164, 165, 165, 170, 176, 179, 198, 214, 218, 221, 225, 233, 238, 241,
       246)
wl <- data.frame(Weight = Weight, Days = Days)
wlmod <- "Weight ~ b0 + b1*2^(-Days/th)"
wlnls <- nls(wlmod, data = wl, start = c(b0 = 1, b1 = 1, th = 1))
summary(wlnls)

##
## Formula: Weight ~ b0 + b1 * 2^(-Days/th)
##
## Parameters:
##    Estimate Std. Error t value Pr(>|t|)
## b0    81.37       2.27    35.9   <2e-16 ***
## b1   102.68       2.08    49.3   <2e-16 ***
## th   141.91       5.29    26.8   <2e-16 ***
## ---
## Signif. codes:  0 '***' 0.001 '**' 0.01 '*' 0.05 '.' 0.1 ' ' 1
##
## Residual standard error: 0.895 on 49 degrees of freedom
##
## Number of iterations to convergence: 6
## Achieved convergence tolerance: 3.28e-06

plot(wl$Days, wl$Weight)
title("Weight loss over time in Days")
points(wl$Days, predict(wlnls), type = "l", col = "red")
```

Despite a very crude set of starting parameters, we see that in this case `nls()` has done a good job of fitting the model to the data.

6.1.2   Regression versus least squares

It is my belief that nls() is mistitled "nonlinear least squares." Its main purpose is to solve nonlinear regression modeling problems, where we are trying to estimate the parameters of models by adjusting them to fit data. In such problems, we rarely get a zero or even a particularly small sum of squares. Moreover, we are often interested in the variability of the parameters, also referred to as *coefficients*, and quantities computed from them. It is worth keeping this in mind when comparing tools for the nonlinear least squares problem. That nonlinear least squares tools are used for nonlinear regression is obvious. That not all nonlinear least squares problems arise from regression applications may be less clear depending on the history and experience of the user.

6.2   A more difficult case

Let us now try an example initially presented by Ratkowsky (1983, p. 88) and developed by Huet et al. (1996, p. 1). This is a model for the regrowth of pasture, with yield as a function of time since the last grazing. Units were not stated for either variable, which I consider bad practice, as units are important information to problems (Nash, 2012). (The criticism can also be made of my own Hobbs Example 1.2. Sometimes the analyst is not given all the information.)

Once again, we set up the computation by putting the data for the problem in a data frame and specifying the formula for the model. This can be as a formula object, but I have found that saving it as a character string seems to give fewer difficulties. Note the ~ (tilde) character that implies "is modeled by" to R. There must be such an element in the formula for nls() as well as for the package **nlmrt**, which we will use shortly.

We also specify two sets of starting parameters, that is, the ones, which is a trivial (but possibly unsuitable) start with all parameters set to 1, and huetstart, which was suggested in Huet et al. (1996) and is provided in the following in the code. However, for reasons that will become clear in the following text, we write the starting vector with the exponential parameters first.

```
options(width = 60)
pastured <- data.frame(time = c(9, 14, 21, 28, 42, 57, 63, 70, 79),
    yield = c(8.93,10.8, 18.59, 22.33, 39.35, 56.11, 61.73, 64.62, 67.08))
regmod <- "yield ~ t1 - t2*exp(-exp(t3+t4*log(time)))"
ones <- c(t3 = 1, t4 = 1, t1 = 1, t2 = 1) # all ones start
huetstart <- c(t3 = 0, t4 = 1, t1 = 70, t2 = 60)
```

Note that the standard nls() of R fails to find a solution from either start.

```
anls <- try(nls(regmod, start = ones, trace = FALSE, data = pastured))
print(strwrap(anls))

## [1] "Error in nlsModel(formula, mf, start, wts) : singular"
```

```
## [2] "gradient matrix at initial parameter estimates"

anlsx <- try(nls(regmod, start = huetstart, trace = FALSE, data = pastured))
print(strwrap(anlsx))

## [1] "Error in nls(regmod, start = huetstart, trace ="
## [2] "FALSE, data = pastured) : singular gradient"
```

Let us now try the routine nlxb() from package **nlmrt**. This function is designed as a replacement for nls(), although it does not cover every feature of the latter. Furthermore, it was specifically prepared as a more robust optimizer than those offered in nls(). nlxb() is more strict than nls() and requires that the problem data be in a data frame. We try both of the suggested starting vectors.

```
require(nlmrt)
anmrt <- nlxb(regmod, start = ones, trace = FALSE, data = pastured)
anmrt

## nlmrt class object: x
## residual sumsquares =   4648.1   on  9 observations
##     after  3    Jacobian and  4 function evaluations
##     name        coeff        SE      tstat     pval        gradient     JSingval
## t3          0.998202        NA        NA       NA       1.889e-08            3
## t4          0.996049        NA        NA       NA       4.15e-08     1.437e-09
## t1          38.8378         NA        NA       NA      -3.04e-11     2.167e-16
## t2          1.00007         NA        NA       NA      -7.748e-10     5.14e-26

anmrtx <- try(nlxb(regmod, start = huetstart, trace = FALSE,
      data = pastured))
anmrtx

## nlmrt class object: x
## residual sumsquares =   8.3759   on  9 observations
##     after  32   Jacobian and  44 function evaluations
##     name        coeff        SE      tstat     pval        gradient     JSingval
## t3          -9.2088      0.8172     -11.27   9.617e-05   5.022e-06        164.8
## t4          2.37778      0.221       10.76   0.0001202  -6.244e-05        3.762
## t1          69.9554      2.363       29.61   8.242e-07   2.196e-07        1.617
## t2          61.6818      3.194       19.31   6.865e-06  -7.295e-07       0.3262
```

The first result has such a large sum of squares that we know right away that there is not a good fit between the model and the data. However, note that the gradient elements are essentially zero, as are three out of four singular values of the Jacobian. The result anmrtx is more satisfactory, although interestingly the gradient elements are not quite as small as those in anmrt. Note that the gradient and Jacobian singular values are NOT related to particular coefficients, but simply displayed in this format for space efficiency.

The Jacobians at the starting points indicate why nls() fails. We use the **nlmrt** function model2jacfun() to generate the R code for the Jacobian, then we compute their singular values. In both cases, the Jacobians could be considered singular, so nls() is unable to proceed.

```
pastjac <- model2jacfun(regmod, ones, funname = "pastjac")
J1 <- pastjac(ones, yield = pastured$yield, time = pastured$time)
svd(J1)$d
```

```
## [1] 3.0000e+00 1.3212e-09 2.1637e-16 3.9400e-26
```

```
J2 <- pastjac(huetstart, yield = pastured$yield, time = pastured$time)
svd(J2)$d
```

```
## [1] 3.0005e+00 1.5143e-01 1.1971e-04 1.8584e-10
```

A useful way to demonstrate the magnitudes of the singular values is to put them in a bar chart (Figure 6.2). If we do this with the singular values from the result anmrtx, we find that the bar for the smallest singular value is very small. Indeed, the lion's share of the variability in the Jacobian is in the first singular value.

```
Jnmrtx <- pastjac(coef(anmrtx), yield = pastured$yield, time = pastured$time)
svals <- svd(Jnmrtx)$d
barplot(svals, main = "Singular values at nlxb solution to pasture problem",
    horiz = TRUE)
```

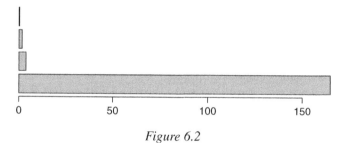

Singular values at nlxb solution to pasture problem

Figure 6.2

The Levenberg–Marquardt stabilization used in nlxb() avoids the singularity of the Jacobian by augmenting its diagonal until it is (computationally) nonsingular. The details of this common approach may be found elsewhere (Nash, 1979, Algorithm 23). The package **minpack.lm** has function nlsLM to implement a similar method and is able to solve the problem from the parameters huetstart, but because it uses an internal function nlsModel() from nls() to set up the computations, it is not able to get started from the values in the starting vector ones.

```
require(minpack.lm)
```

```
## Loading required package: minpack.lm
```

```
aminp <- try(nlsLM(regmod, start = ones, trace = FALSE, data = pastured))
```

```
summary(aminp)

##    Length     Class      Mode
##         1 try-error character

aminpx <- try(nlsLM(regmod, start = huetstart, trace = FALSE,
    data = pastured))
print(aminpx)

## Nonlinear regression model
##    model: yield ~ t1 - t2 * exp(-exp(t3 + t4 * log(time)))
##    data: pastured
##    t3     t4     t1     t2
## -9.21   2.38  69.96  61.68
##    residual sum-of-squares: 8.38
##
## Number of iterations to convergence: 42
## Achieved convergence tolerance: 1.49e-08
```

Both **nlmrt** and **minpack.lm** have functions to deal with a model expressed as an R function rather than an expression, and **nlmrt** also has tools to convert an expression to a function. The package **pracma** has a similar function lsqnonlin(). CRAN package **MarqLevAlg** is also intended to solve unconstrained nonlinear least squares problems but works with the sum-of-squares function rather than the residuals.

There are some other tools for R that aim to solve nonlinear least squares problems, to the extent that there are two packages called **nls2**. At the time of writing (August 2013), we have not yet been able to successfully use the INRA package nls2. This is a quite complicated package and is not installable as a regular R package using install.packages(). The **nls2** package on CRAN by Gabor Grothendieck is intended to combine a grid search with the nls() function. The following is an example of its use to solve the pasture regrowth problem starting at ones:

```
require(nls2)

## Loading required package: nls2
## Loading required package: proto

set.seed(123) # for reproducibility
regmodf <- as.formula(regmod) # just in case
m100 <- c(t1 = -100, t2 = -100, t3 = -100, t4 = -100)
p100 <- (-1) * m100
gstart <- data.frame(rbind(m100, p100))
anls2 <- try(nls2(regmodf, start = gstart, data = pastured, algorithm =
"random-search", control = list(maxiter = 1000)))
print(anls2)

## Nonlinear regression model
##    model: yield ~ t1 - t2 * exp(-exp(t3 + t4 * log(time)))
##    data: pastured
```

```
##     t1     t2     t3     t4
##   11.6  -44.8   84.8  -26.1
##   residual sum-of-squares: 1377
##
## Number of iterations to convergence: 1000
## Achieved convergence tolerance: NA
```

This is not nearly so good as our Marquardt solutions from `nlxb()` or `nlsLM()`, but trials with those routines from other starting points give "answers" that are all over the place. Something is not right!

We could, however, try the partial linearity tools in `nls2()`. We note that the parameter *t*1 is an intercept and *t*2 multiplies the exponential construct involving *t*3 and *t*4, so we can treat the problem as one of finding the latter two parameters with a nonlinear modeling tool and solve for the other two with linear regression. Package **nlmrt** does not yet have such a capability, but **nls2** does. The following code shows how to use it.

```
require(nls2)
set.seed(123) # for reproducibility
plinform <- yield ~ cbind(1, -exp(-exp(t3 + t4 * log(time))))
gstartpl <- data.frame(rbind(c(-10, 1), c(10, 8)))
names(gstartpl) <- c("t3", "t4")
anls2plb <- try(nls2(plinform, start = gstartpl, data = pastured,
    algorithm = "plinear-brute", control = list(maxiter = 200)))
print(anls2plb)

## Nonlinear regression model
##   model: yield ~ cbind(1, -exp(-exp(t3 + t4 * log(time))))
##    data: pastured
##    t3     t4 .lin1 .lin2
## -10.0    2.5  75.5  64.8
##   residual sum-of-squares: 38.6
##
## Number of iterations to convergence: 225
## Achieved convergence tolerance: NA

## ===================================
anls2plr <- try(nls2(plinform, start = gstartpl, data = pastured,
    algorithm = "plinear-random", control = list(maxiter = 200)))
print(anls2plr)

## Nonlinear regression model
##   model: yield ~ cbind(1, -exp(-exp(t3 + t4 * log(time))))
##    data: pastured
##    t3     t4 .lin1 .lin2
## -6.42   1.57 87.95 84.82
##   residual sum-of-squares: 29.9
##
## Number of iterations to convergence: 200
## Achieved convergence tolerance: NA
```

Unfortunately, we see two rather different results, suggesting that something is not satisfactory with this model. However, we can use the final results of **nls2** and try them with nls(). Here we now take advantage of putting *t3* and *t4* first in the starting vector, because we only need to fix the parameter names. From both these starts, we find the same solution as nlxb().

To visually compare the models found from each of the **nls2** solutions above with the best solution found, we graph the fitted lines for all three models. So that we can compute the fitted values and graph them; we will use the wrapnls() function of **nlmrt** and graph the fits versus the original data (Figure 6.3). wrapnls() simply calls nls() with the results parameters of nlxb() to obtain the special result structure nls() offers. The residuals can also be displayed in boxplots to compare their sizes, and we present that display also (Figure 6.4).

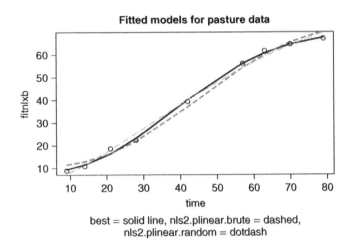

best = solid line, nls2.plinear.brute = dashed,
nls2.plinear.random = dotdash

Figure 6.3

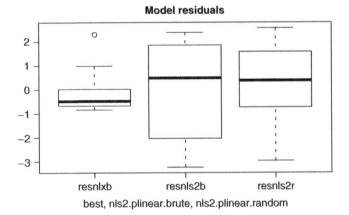

best, nls2.plinear.brute, nls2.plinear.random

Figure 6.4

```
require(nlmrt)

## Loading required package: nlmrt

splb <- coef(anls2plb)
splr <- coef(anls2plr)
names(splb) <- names(huetstart)
names(splr) <- names(huetstart)
anlsfromplb <- nls(regmod, start = splb, trace = FALSE, data = pastured)
anlsfromplr <- nls(regmod, start = splr, trace = FALSE, data = pastured)
agomp <- wrapnls(regmodf, start = huetstart, data = pastured)
fitnlxb <- fitted(agomp)
fitnls2b <- fitted(anls2plb)
fitnls2r <- fitted(anls2plr)
plot(pastured$time, fitnlxb, type = "l", lwd = 2, xlab = "time")
points(pastured$time, fitnls2b, col = "red", type = "l", lty = "dashed",
    lwd = 2)
points(pastured$time, fitnls2r, col = "blue", type = "l", lty = "dotdash",
    lwd = 2)
points(pastured$time, pastured$yield)
title(main = "Fitted models for pasture data")
sstr <- "best = solid line, nls2.plinear.brute = dashed,
  nls2.plinear.random = dotdash"
title(sub = sstr, cex.sub = 0.8)
```

```
resnlxb <- pastured$yield - fitnlxb
resnls2b <- pastured$yield - fitnls2b
resnls2r <- pastured$yield - fitnls2r
```

While the graphs show the nlxb() solution is "better" than the others, there remains a sense of discomfort that a reasonable solution has not been reliably obtained. Other experiments show that nlxb() can give **unsatisfactory** solutions from many sets of starting values. We were lucky here. Note that nls2() must be provided with intervals that include "good" parameters, thereby requiring us to have an idea where these are.

A similar variation in results is found by using the **nlmrt** codes model2ssfun() and model2grfun() to generate an objective function and its gradient for use with optimization routines.

```
ssfn <- model2ssfun(regmodf, ones)
grfn <- model2grfun(regmodf, ones)
time <- pastured$time
yield <- pastured$yield
# aopto<-optim(ones, ssfn, gr=grfn, yield=yield, time=time,
# method='BFGS', control=list(maxit=10000)) print(aopto) ##
# We did not run this one.
aopth <- optim(huetstart, ssfn, gr = grfn, yield = yield, time = time,
    method = "BFGS", control = list(maxit = 10000))
print(aopth)

## $par
##      t3      t4      t1      t2
## -9.2089  2.3778 69.9552 61.6815
```

```
##
## $value
## [1] 8.3759
##
## $counts
## function gradient
##       92       36
##
## $convergence
## [1] 0
##
## $message
## NULL
```

The conclusion of all this? While we can find acceptable solutions, the tools may not automatically provide us the evidence that we have the BEST solution. This is an area that is still worthy of research (into the mathematics) and development (of the software tools).

6.3   The structure of the `nls()` solution

The R function `str()` can quickly reveal what is returned by `nls()`, and it is an impressive set of capabilities. As indicated, my only concern is that it is too rich for many applications and users. Some users will also find that the nomenclature is at odds with usage in the optimization community. Suppose the solutions to the weight loss problem in Section 6.1.1 from `nls()` and `nlxb()` are `wlnls` and `wlnlxb`, respectively. The Jacobian of the residuals at the solution is obtained from `wlnls` as `wlnls$m$gradient()` (note that it is a function call), while from `wlnlxb` as `wlnlxb$jacobian`. This use of the term "gradient" carries over to the error message "singular gradient," which, I believe, confuses users.

```
require(nlmrt)
wlmod <- "Weight ~ b0 + b1*2^(-Days/th)"
wlnlxb <- nlxb(wlmod, data = wl, start = c(b0 = 1, b1 = 1, th = 1))
wlnlxb

## nlmrt class object: x
## residual sumsquares =  39.245  on  52 observations
##     after  9    Jacobian and  13 function evaluations
##    name       coeff       SE      tstat     pval     gradient    JSingval
## b0          81.3751     2.266     35.9        0    7.707e-05        8.53
## b1          102.683      2.08    49.36        0   -8.111e-05       1.427
## th          141.908     5.289    26.83        0   -2.634e-05      0.1471
```

Both `nlxb()` and `nls()` can use a function `coef()` to get at the parameters or coefficients of the linear model. Thus,

```
coef(wlnls)
```

```
##       b0      b1      th
## 81.374 102.684 141.910
```

```
coef(wlnlxb)
```

However, internally, `nlxb()` returns the coefficients as `wlnlxb$coeffi-`
`cients`, that is, as a simple element of the solution. By contrast, one needs to
extract them from the `nls()` solution object using `wlnls$m$getPars()`. This
is a function that appears to be undocumented by the conventional R approach of
using `?getPars` or `??getPars`. It is visible by `str(wlnls)`. There is also
documentation in a web page titled "Create an nlsModel Object" at http://web.njit.
edu/all_topics/Prog_Lang_Docs/html/library/nls/html/nlsModel.html. However, I
could not find this page in the main R manuals on the www.r-project.org web site.
My complaint is not that the information is at all difficult to obtain once the mechanism is known but that finding how to do so requires more digging than it should.

Certainly, as I point out in the Chapter 3, it is difficult to structure the output
of complicated tools like `nls()` in a manner that is universally convenient.
Knowing how to use functions `getPars()` or `coef(wlnls)`, users can
quickly access the information they need. There is yet another possibility in
`summary(wlnls)$coefficients[,1]`.

It is helpful to be able to extract a set of parameters (coefficients) from one
solution or model and try it in another solver as a crude check for stability of these
estimates or, as in the control parameter `follow.on` in package **optimx**, to be
able to run an abbreviated solution process with one solver to get close to a solution
and then use another to complete the process. This approach tries to take advantage
of differential advantages of different methods.

Unlike `nls()` or `nlsLM` of **minpack.lm**, **nlmrt** tools generally do not
return the large nls-style object. In part, this is not only because the **nlmrt** tools
are intended to deal with the zero-residual case where some of the parts of the
`nlsModel` structure make little sense but also because I prefer a simpler output
object. However, as mentioned, there is `wrapnls` that calls `nls()` appropriately
(i.e., with bounds) after running `nlxb()`.

6.4 Concerns with `nls()`

`nls()` seems to be designed mainly for nonlinear regression, so it is not entirely
suitable for general nonlinear least squares computations. In my view, the following
are the main issues, which are sometimes seen in similar codes both in R and other
systems.

- It is specifically NOT indicated for problems where the residuals are small or zero. Thus it excludes many approximation problems.

- It frequently fails to find a solution to problems that other methods solve easily, giving a cryptic "singular gradient" error message.

- Bounds on parameters require the use of the "port" algorithm. Unless this is specified, bounds are ignored.

- Because the solution object from nls() is complicated, extracting information from it can be error prone. Some of this output (e.g., "achieved convergence tolerance") is not fully explained.

6.4.1   Small residuals

The nls() manual is specific in warning that it should not be used for problems with "artificial 'zero-residual' data." In such cases, the manual suggests adding some noise to the data. As we have seen, nonlinear equations problems give rise to zero-residual nonlinear least squares problems, as do certain approximation and interpolation problems, where adding noise destroys the value of the solutions. The restriction with nls() arises because of software design choices rather than fundamental algorithmic reasons. While the implementation choices for nls() may be sensible in the context of traditional statistical estimation, they are unhelpful to general users.

In fact, the nonlinear least squares problem is generally well defined when the residuals are small or zero. The difficulty for nls() comes not from the Gauss–Newton method, but rather the particular relative offset convergence test (Bates and Watts, 1981). This tries to compare the change in parameters to a measure that is related to the size of the residuals, essentially dividing zero by zero. In an appropriate context, the Bates–Watts convergence test is a "very good idea," and I decided to incorporate a version of this test in **nlmrt** while writing this section. However, far from all nonlinear least squares problems fit its assumptions. I am, in fact, surprised nls() does not appear to offer a way to simply select a more mundane convergence test.

Packages **nlmrt** and **minpack.lm** have ways to select suitable termination criteria when the residuals are very small. In **nlmrt**, the functions nlxb() and nlfb() allow the control list elements rofftest and smallsstest to be set FALSE. The first choice turns off the relative offset test, thereby trying harder to find a minimum. The second choice ignores a test of the size of the sum of squares. If we expect our residuals to be somewhere in the scale of 1 to 10, a very small sum of squares would imply we are finished, and keeping the default test for a small sum of squares makes sense. For problems where our fit could be exact, we should turn off this termination condition.

```
x <- 1:10
y <- 2 * x + 3 # perfect fit
yeps <- y + rnorm(length(y), sd = 0.01) # added noise
anoise <- nls(yeps ~ a + b * x, start = list(a = 0.12345, b = 0.54321))
summary(anoise)

##
## Formula: yeps ~ a + b * x
##
## Parameters:
##    Estimate Std. Error t value Pr(>|t|)
## a  3.01098    0.00699     431  <2e-16 ***
## b  1.99857    0.00113    1775  <2e-16 ***
## ---
## Signif. codes:  0 '***' 0.001 '**' 0.01 '*' 0.05 '.' 0.1 ' ' 1
##
## Residual standard error: 0.0102 on 8 degrees of freedom
##
## Number of iterations to convergence: 2
## Achieved convergence tolerance: 3.91e-09

aperf <- try(nls(y ~ a + b * x, start = list(a = 0.12345, b = 0.54321)))
print(strwrap(aperf))

## [1] "Error in nls(y ~ a + b * x, start = list(a = 0.12345,"
## [2] "b = 0.54321)) : number of iterations exceeded maximum"
## [3] "of 50"

ldata <- data.frame(x = x, y = y)
aperfn <- try(nlxb(y ~ a + b * x, start = list(a = 0.12345, b = 0.54321),
    data = ldata))
aperfn

## nlmrt class object: x
## residual sumsquares =  4.5342e-20  on  10 observations
##     after  3    Jacobian and  4 function evaluations
##   name       coeff        SE        tstat      pval     gradient    JSingval
## a               3   5.143e-11  5.833e+10         0    1.671e-10       19.82
## b               2   8.289e-12  2.413e+11         0   -9.542e-10       1.449
```

6.4.2   Robustness – "singular gradient" woes

We have already seen an example of nls() stopping with the "singular gradient" termination message. While Marquardt approaches may also fail, they are generally more robust than the Gauss–Newton method of nls().

The following problem was posted on the Rhelp list. It has only four observations to try to fit three parameters, so we may fully expect strange results. The first try results in the common "singular gradient" error with nls(), but nlxb() can find a solution.

```
x <- c(60, 80, 100, 120)
y <- c(0.8, 6.5, 20.5, 45.9)
mydata <- data.frame(x, y)
pnames <- c("a", "b", "d")
npar <- length(pnames)
st <- c(1, 1, 1)
names(st) <- pnames
rnls <- try(nls(y ~ exp(a + b * x) + d, start = st, data = mydata),
    silent = TRUE)
if (class(rnls) == "try-error") {
    cat("nls() failed (singular gradient?)\n")
} else {
    summary(rnls)
}

## nls() failed (singular gradient?)

require(nlmrt)
rnlx0 <- try(nlxb(y ~ exp(a + b * x) + d, start = st, data = mydata),
    silent = TRUE)
cat("Found sum of squares ", rnlx0$ssquares, "\n")

## Found sum of squares  1.701e+104

print(rnlx0$coefficients)

##              a           b           d
## 5.0003e-01  9.9583e-01 -5.9072e+47

## Now turn off relative offset test
rnlx1 <- try(nlxb(y ~ exp(a + b * x) + d, start = st, data = mydata,
    control = list(rofftest = FALSE)), silent = TRUE)
cat("Found sum of squares ", rnlx1$ssquares, "\n")

## Found sum of squares  2.3892e+50

coef(rnlx1)
## And the sumsquares size test
rnlx2 <- try(nlxb(y ~ exp(a + b * x) + d, start = st, data = mydata,
    control = list(rofftest = FALSE, smallsstest = FALSE)), silent = TRUE)
cat("Found sum of squares ", rnlx2$ssquares, "\n")

## Found sum of squares  0.57429

coef(rnlx2)
```

Note how wild the results are unless we are very aggressive in keeping the algorithm working by turning off relative offset and "small" sumsquares tests. In this case, "small" is $(100*.Machine\$double.eps)^4 = 2.430865e-55$ times the initial sum of squares (of the order of 1e+105), which is still bigger than 1e+50. This is a very extreme problem, although it looks quite ordinary.

To explore the possibility that different starting values for the parameters might be important for success, we will generate 1000 sets of starting parameters in a box from 0 to 8 on each edge using uniformly distributed pseudorandom numbers.

Then we will try to solve the problem just presented from these starting points with nls() and nlxb() (from **nlmrt**). We will not show the code here, which is simple but a bit tedious. Note that the nlxb() is called with the control elements rofftest=FALSE and smallsstest=FALSE. Besides actual failures to terminate normally, we judge any solution with a sum of squares exceeding 0.6 as failures.

```
## Proportion of nls  runs that failed =  0.728
## Proportion of nlxb runs that failed =  0
## Average iterations for nls = 20.018
## Average jacobians for nlxb = 66.684
## Average residuals for nlxb = 80.463
```

This is just one numerical experiment – a very extreme one – and we should be cautious about the lessons we take from it. However, it does mirror my experiences, namely,

- the Marquardt approach is generally much more successful in finding solutions than the Gauss–Newton approach that nls() uses;

- a relative offset test of the type always used in nls() but which can be ignored in nlxb() is not appropriate for extreme problems;

- when they do find solutions, Gauss–Newton methods like nls() are generally more efficient than Marquardt methods.

6.4.3   Bounds with nls()

My concerns here are best summarized by quoting two portions of the nls() manual:

```
Bounds can only be used with the "port" algorithm.
They are ignored, with a warning, if given for other algorithms.

The 'algorithm = "port"' code appears unfinished, and does not
  even check that the starting value is within the bounds.  Use with
  caution, especially where bounds are supplied.
```

In fact, I have NOT found any issues with using the "port" code. I am more concerned that bounds are ignored, although a warning is issued. I would rather that the program stop in this case, as users are inclined to ignore warnings.

The nlxb() function handles bounds and seems to work properly. I have found (and reported to the maintainers) that nlsLM() can sometimes stop too soon. I believe this is due to a failure to project the Marquardt search direction onto the feasible hyperplane, which requires action inside the Marquardt code. In actuality, one needs to check if the search will or will not bump into the bound and either free or impose the constraint. In the case of package **minpack.lm**, the program code where this must happen is inside non-R code that is messy to adjust.

Using the scaled Hobbs problem of Section 1.2, let us try simple upper bounds at 2 on all parameters and apply nls(), nlxb(), and nlsLM(). We need to ensure that we consider the sum of squares, which shows the last one has stopped prematurely.

```
ydat <- c(5.308, 7.24, 9.638, 12.866, 17.069, 23.192, 31.443,
    38.558, 50.156, 62.948, 75.995, 91.972) # for testing
tdat <- 1:length(ydat) # for testing
weeddata <- data.frame(y = ydat, t = tdat)
require(nlmrt)
require(minpack.lm)
hobmod <- "y~100*b1/(1+10*b2*exp(-0.1*b3*t))"
st <- c(b1 = 1, b2 = 1, b3 = 1)
low <- c(-Inf, -Inf, -Inf)
up <- c(2, 2, 2)


cat("try nlxb\n")

## try nlxb

anlxb2 <- nlxb(hobmod, st, data = weeddata, lower = low, upper = up)
anlxb2

## nlmrt class object: x
## residual sumsquares =  839.73  on  12 observations
##     after  7    Jacobian and  8 function evaluations
##   name       coeff        SE       tstat      pval      gradient    JSingval
## b1            2U          NA        NA        NA        0           60.46
## b2          1.78636       NA        NA        NA       -3.637e-05    0
## b3            2U          NA        NA        NA        0           0

try(anls2p <- nls(hobmod, st, data = weeddata, lower = low, upper = up,
    algorithm = "port"))
summary(anls2p)

##
## Formula: y ~ 100 * b1/(1 + 10 * b2 * exp(-0.1 * b3 * t))
##
## Parameters:
##    Estimate Std. Error t value Pr(>|t|)
## b1    2.00      3.96     0.50    0.63
## b2    1.79      3.08     0.58    0.58
## b3    2.00      1.21     1.66    0.13
##
## Residual standard error: 9.66 on 9 degrees of freedom
##
## Algorithm "port", convergence message: relative convergence (4)

anls2p$m$deviance()

## [1] 839.73

aLM2 <- nlsLM(hobmod, st, data = weeddata, lower = low, upper = up)
summary(aLM2)

##
## Formula: y ~ 100 * b1/(1 + 10 * b2 * exp(-0.1 * b3 * t))
```

```
##
## Parameters:
##     Estimate Std. Error t value Pr(>|t|)
## b1     2.00      5.12     0.39    0.71
## b2     2.00      4.49     0.44    0.67
## b3     2.00      1.41     1.42    0.19
##
## Residual standard error: 10.5 on 9 degrees of freedom
##
## Number of iterations to convergence: 2
## Achieved convergence tolerance: 1.49e-08

aLM2$m$deviance()

## [1] 987.58
```

6.5   Some ancillary tools for nonlinear least squares

Because of the importance and generality of nonlinear least squares problems, there are **many** add-on tools. We consider only some of them here.

6.5.1   Starting values and self-starting problems

Note that all of the methods we discuss assume that the user has in some way provided some data, and most approaches also ask for some preliminary guesses for the values of the parameters (coefficients) of the model.

nlxb() always requires a start vector of parameters. nls() includes a provision to try 1 for all parameters as a start if none is provided. While I have considered making the change to nlxb() to do this and often use such a starting vector myself, I believe it should be a conscious choice. Therefore, I am loath to provide such a default.

Good starting vectors really do help to save effort and get better solutions. For nls(), there are a number of **self-starting** models. In particular, the SSlogis model provides a form of the three-parameter logistic that we used to model the Hobbs data earlier. We explore this reparameterization in Chapter 16, and the details of the model are there, but it is useful to show how to use the tool that is in R.

```
## Put in weeddata here as precaution. Maybe reset workspace.
anlss2 <- nls(y ~ SSlogis(t, p1, p2, p3), data = weeddata)
summary(anlss2)

##
## Formula: y ~ SSlogis(t, p1, p2, p3)
##
## Parameters:
##     Estimate Std. Error t value Pr(>|t|)
## p1 196.1863    11.3069    17.4  3.2e-08 ***
## p2  12.4173     0.3346    37.1  3.7e-11 ***
## p3   3.1891     0.0698    45.7  5.8e-12 ***
## ---
```

```
## Signif. codes:  0 '***' 0.001 '**' 0.01 '*' 0.05 '.' 0.1 ' ' 1
##
## Residual standard error: 0.536 on 9 degrees of freedom
##
## Number of iterations to convergence: 0
## Achieved convergence tolerance: 2.13e-07

anlss2$m$deviance()

## [1] 2.587
```

Note that the sum of squared residuals is the square of the residual standard error times the degrees of freedom. We can also use the `deviance()` supplied in the solution object.

6.5.2   Converting model expressions to sum-of-squares functions

Sometimes we want to use an R function to compute the residuals, Jacobian, or the sum-of-squares function so we can use optimization tools like those in `optim()`. See Section 6.6. So far, we have looked at tools that express the problem in terms of a model **expression**.

We can, of course, write such functions manually, but package **nlmrt** and some others offer tools to do this, such as four tools that start with

- `model2ssfun()`: to create a sum of squares function;

- `model2resfun()`: to create a function that computes the residuals;

- `model2jacfun()`: to create a function that computes the Jacobian;

- `model2grfun()`: to create a function that computes the gradient.

These functions require the `formula` as well as a "typical" starting vector (to provide the names of the parameters).

There are also two tools that work with the residual and Jacobian functions and compute the sum of squares and gradient from them.

- `modss()`: to compute the sum of squares by computing the residuals and forming their cross-product;

- `modgr()`: to compute the gradient of the sum of squares from the residuals and the Jacobian.

These two functions require functions for the residuals and the Jacobian in their calling syntax.

6.5.3   Help for nonlinear regression

This book is mostly about the engines inside the vehicles for nonlinear optimization, but many people are more concerned with the bodywork and passenger features of the nonlinear modeling cars. Fortunately, there are some useful reference works on this topic, and some are directed at R users. In particular, Ritz and Streibig (2008) is helpful. There are also some good examples in Huet et al. (1996), but the software used is not mainstream to R. Older books have a lot of useful material, but their ideas must be translated to the R context. Among these are Bates and Watts (1988), Seber and Wild (1989), Nash and Walker-Smith (1987), Ratkowsky (1983), and Ross (1990).

6.6   Minimizing R functions that compute sums of squares

Package **nlmrt** and package **minpack.lm** include tools to minimize sums-of-squares residuals expressed as functions rather than expressions. They both use Marquardt-based methods to solve nonlinear least squares problems. We have already seen nlxb() and nlsLM(). However, both nls.lm from **minpack.lm** and nlfb from **nlmrt** are designed to work with a **function** to compute the vector of residuals at a given set of parameters, with (optionally) a function to compute the Jacobian matrix.

```
## weighted nonlinear regression
Treated <- as.data.frame(Puromycin[Puromycin$state == "treated",
    ])

weighted.MM <- function(resp, conc, Vm, K) {
    ## Purpose: exactly as white book p. 451 -- RHS for nls()
    ## Weighted version of Michaelis-Menten model
    ## -----------------------------------------------------------
    ## Arguments: 'y', 'x' and the two parameters (see book)
    ## -----------------------------------------------------------
    ## Author: Martin Maechler, Date: 23 Mar 2001

    pred <- (Vm * conc)/(K + conc)
    (resp - pred)/sqrt(pred)
}

wtMM <- function(x, resp, conc) {
    # redefined for nlfb()
    Vm <- x[1]
    K <- x[2]
    res <- weighted.MM(resp, conc, Vm, K)
}

start <- list(Vm = 200, K = 0.1)

Pur.wt <- nls(~weighted.MM(rate, conc, Vm, K), data = Treated,
    start)
print(summary(Pur.wt))

##
## Formula: 0 ~ weighted.MM(rate, conc, Vm, K)
##
```

```
## Parameters:
##    Estimate Std. Error t value Pr(>|t|)
## Vm 2.07e+02   9.22e+00   22.42  7.0e-10 ***
## K  5.46e-02   7.98e-03    6.84  4.5e-05 ***
## ---
## Signif. codes:  0 '***' 0.001 '**' 0.01 '*' 0.05 '.' 0.1 ' ' 1
##
## Residual standard error: 1.21 on 10 degrees of freedom
##
## Number of iterations to convergence: 5
## Achieved convergence tolerance: 3.83e-06
```

```
require(nlmrt)
anlf <- nlfb(start, resfn = wtMM, jacfn = NULL, trace = FALSE,
    conc = Treated$conc, resp = Treated$rate)
anlf
```

```
## nlmrt class object: x
## residual sumsquares =  14.597  on  12 observations
##     after  6    Jacobian and  7 function evaluations
##    name      coeff        SE      tstat       pval     gradient   JSingval
## Vm         206.835      9.225      22.42   7.003e-10  -9.708e-11    230.7
## K          0.0546112   0.007979    6.845   4.488e-05  -7.255e-09    0.131
```

6.7   Choosing an approach

We have seen that solving nonlinear least squares problems with R generally takes one of two approaches, either via a modeling **expression** or one or more **functions** that compute the residuals and related quantities. These approaches have a quite different "feel." When we work with residual functions, we will often think of their derivatives – the Jacobian – whereas expression-based tools attempt to generate such information. In that task, nls() has more features than nlxb(), as we shall see in the following example. By contrast, we generally have a lot more flexibility in defining the residuals when taking a function approach.

The problem we shall use to illustrate the differences in approach will be referred to as the "two straight line" problem. Here we want to model data using two straight line models that intersect at some point (Figure 6.5). This problem was of interest to a colleague at Agriculture Canada in the 1970s. See Nash and Price (1979). For the present example, we will generate some data.

```
set.seed(1235)
x <- 1:40
xint <- 20.5 * rep(1, length(x))
sla <- 0.5
slb <- -0.5
yint <- 30
idx <- which(x <= xint)
ymod <- {
    yint + (x - xint) * slb
}
ymod[idx] <- yint + (x[idx] - xint[idx]) * sla
ydata <- ymod + rnorm(length(x), 0, 2)
```

```
plot(x, ymod, type = "l", ylim = c(0, 40), ylab = "y")
points(x, ydata)
title(main = "2 straight lines data")
title(sub = "Lines are those used to generate data")
```

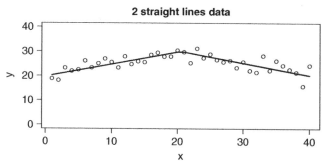

2 straight lines data

Lines are those used to generate data

Figure 6.5

We can model this data quite nicely with `nls()` if we take a little care in how we define the modeling expression and select starting values. Note that we are supposed to be able to leave out `start`, but that does not seem to work here. Nor does putting in a very naive set of starting parameters. However, a very rough choice does work.

```
tslexp <- "y ~ (yint + (x-xint)*slb)*(x >= xint) + (yint + (x-xint)*sla)*
( x < xint)"
mydf <- data.frame(x = x, y = ydata)
mnls <- try(nls(tslexp, trace = FALSE, data = mydf))
strwrap(mnls)

## [1] "Error in getInitial.default(func, data, mCall ="
## [2] "as.list(match.call(func, : no 'getInitial' method"
## [3] "found for \"function\" objects"

mystart <- c(xint = 1, yint = 1, sla = 1, slb = 1)
mnls <- try(nls(tslexp, start = mystart, trace = FALSE, data = mydf))
strwrap(mnls)

## [1] "Error in nlsModel(formula, mf, start, wts) : singular"
## [2] "gradient matrix at initial parameter estimates"

myst2 <- c(xint = 15, yint = 25, sla = 1, slb = -1)
mnls2 <- try(nls(tslexp, start = myst2, trace = FALSE, data = mydf))
summary(mnls2)

##
## Formula: y ~ (yint + (x - xint) * slb) * (x >= xint) + (yint + (x - xint) *
##      sla) * (x < xint)
##
```

```
## Parameters:
##       Estimate Std. Error t value Pr(>|t|)
## xint  19.9341     1.5623   12.76  6.4e-15 ***
## yint  29.4904     0.6922   42.61  < 2e-16 ***
## sla    0.4617     0.0915    5.05  1.3e-05 ***
## slb   -0.4276     0.0787   -5.43  4.0e-06 ***
## ---
## Signif. codes:  0 '***' 0.001 '**' 0.01 '*' 0.05 '.' 0.1 ' ' 1
##
## Residual standard error: 2.18 on 36 degrees of freedom
##
## Number of iterations to convergence: 5
## Achieved convergence tolerance: 8.05e-09

mnls2$m$deviance()

## [1] 171.62
```

Unfortunately, as yet `nlxb()` does not handle this model form.

```
require(nlmrt)
mnlxb <- try(nlxb(tslexp, start = mystart, trace = FALSE, data = mydf))
strwrap(mnlxb)

## [1] "Error in deriv.default(parse(text = resexp),"
## [2] "names(start)) : Function ''<'' is not in the"
## [3] "derivatives table"
```

By the time this book is published, this obstacle may be overcome. However, in the meantime, it is relatively straightforward to convert the problem to one involving a residual function and to call `nlfb()` using a numerical Jacobian approximation.

```
myres <- model2resfun(tslexp, mystart)
## Now check the calculation
tres <- myres(myst2, x = mydf$x, y = mydf$y)
tres

##  [1]  -7.85402  -6.18029 -10.22992  -7.97355  -7.47842
##  [6] -10.14639  -6.34569  -7.05972  -7.98057  -5.55515
## [11]  -2.36927  -5.91993  -1.74969  -1.90661  -0.60365
## [16]  -4.57540  -6.47660  -5.84212  -6.91754 -10.38309
## [21] -10.74823  -7.26064 -14.16750 -11.29679 -13.85938
## [26] -12.55585 -12.68729 -14.22140 -12.35207 -15.71322
## [31] -13.14478 -13.50182 -21.21353 -16.25285 -21.17923
## [36] -20.28866 -19.61445 -19.45144 -14.75980 -24.25293

## But jacobian fails
jactry <- try(myjac <- model2jacfun(tslexp, mystart))
strwrap(jactry)

## [1] "Error in deriv.default(parse(text = resexp), pnames)"
## [2] ": Function ''<'' is not in the derivatives table"
```

```
## Try nlfb with no jacfn.
mnlfb <- try(nlfb(mystart, resfn = myres, trace = FALSE, x = mydf$x,
    y = mydf$y))
mnlfb

## nlmrt class object: x
## residual sumsquares =   434.73   on  40 observations
##     after  4    Jacobian and  5 function evaluations
##   name         coeff        SE       tstat      pval       gradient    JSingval
## xint        -10.4204        NA         NA        NA       9.993e-09      208.8
## yint         25.1081        NA         NA        NA       7.067e-10      2.211
## sla                1        NA         NA        NA               0     7.68e-09
## slb     -0.00182172         NA         NA        NA       3.604e-08     5.527e-23

## Also try from better start
mnlfb2 <- try(nlfb(myst2, resfn = myres, trace = FALSE, x = mydf$x,
    y = mydf$y))
mnlfb2

## nlmrt class object: x
## residual sumsquares =   171.62   on  40 observations
##     after  6    Jacobian and  7 function evaluations
##   name         coeff        SE       tstat       pval        gradient    JSingval
## xint         19.9341       1.562      12.76    6.217e-15    -2.978e-13        54
## yint         29.4904      0.6922      42.61            0     6.917e-12     49.62
## sla         0.461741     0.09145      5.049    1.294e-05     3.462e-11     3.148
## slb        -0.427626     0.07868     -5.435     3.96e-06    -3.232e-11     1.395

## It will also work with mnlfb2j<-try(nlfb(myst2,
## resfn=myres, jacfn=NULL, trace=FALSE, x=mydf$x, y=mydf$y))
## or with Package pracma's function lsqnonlin (but there is
## not a good summary())
require(pracma)
alsqn <- lsqnonlin(myres, x0 = myst2, x = mydf$x, y = mydf$y)
print(alsqn$ssq)

## [1] 171.62

print(alsqn$x)

##      xint      yint       sla       slb
## 19.93411 29.49038  0.46174 -0.42763
```

We can also show that **minpack.lm** can use either its expression-based nlsLM() or function-based nls.lm() to solve this problem.

It is also possible to think of the problem as a one-dimensional minimization, where we choose an *x*-value for the intersection of the two lines and compute separate linear models to the left and right of this divider, with the objective function the sum of the two sums of squares. The glitch in this approach (which "mostly" works) is that sometimes the two lines do not intersect near the chosen *x*-value. In such cases, the apparent sum of squares is too low.

This problem is one of many where a relatively simple change to a linear modeling task gives rise to a problem that is in some ways nonlinear. Many constrained linear model problems, such as nonnegative least squares, the present two or more straight lines problem, and the Hassan problem from Nash (1979, p. 184) that is treated in Chapter 13 can all be approached using nonlinear least squares tools.

6.8 Separable sums of squares problems

In many situations, we encounter problems where we need to solve a nonlinear least squares problem but the model to be fitted has several linear parameters, or else can be thus reorganized. We already encountered this in the Pasture problem earlier. A classic example is the multiple exponentials model that occurs when we have, for example, the radioactive decay over time of a mixture of two radioactive elements with different half-lives. The measured radioactivity y is the sum of the radiation from each and will decay over time, which is given the variable x here.

$$y_t = c_1 * \exp(-a_1 x) + c_2 * \exp(-a_2 x) \tag{6.1}$$

where c_1 and c_2 are the amounts or concentrations of each of the elements and the a_1 and a_2 are their decay rates.

In practice, it is fairly easy to get a good "fit" to the data, but if we artificially generate data from a particular example of such a model, it is exceedingly difficult to recover the input parameters. There are often many parameter sets that fit the data equally well. This has been well known for a long time, and we will use a classic example from Lanczos (1956, p. 276). This is two-decimal data for 24 values of the x, y data generated from a sum of three exponentials.

```
require(NISTnls)
L1 <- Lanczos1 # The data frame of data, but more than 2 decimals on y
L1[, "y"] <- round(L1[, "y"], 2)
L1
```

```
##          y    x
## 1     2.51 0.00
## 2     2.04 0.05
## 3     1.67 0.10
## 4     1.37 0.15
## 5     1.12 0.20
## 6     0.93 0.25
## 7     0.77 0.30
## 8     0.64 0.35
## 9     0.53 0.40
## 10    0.45 0.45
## 11    0.38 0.50
## 12    0.32 0.55
## 13    0.27 0.60
## 14    0.23 0.65
## 15    0.20 0.70
## 16    0.17 0.75
## 17    0.15 0.80
## 18    0.13 0.85
## 19    0.11 0.90
## 20    0.10 0.95
```

```
## 21 0.09 1.00
## 22 0.08 1.05
## 23 0.07 1.10
## 24 0.06 1.15
```

It is straightforward to write down a function for the residuals (here in "model-data" form). The Jacobian function is also straightforward. We also set some starting vectors, for which parameter names are provided. This is not strictly necessary for a function approach to the problem but is required for the expression approach of `nls()`. We write the model expression at the end of the code segment.

```
lanc.res <- function(b, data) {
    # 3 exponentials
    res <- rep(NA, length(data$x))
    res <- b[4] * exp(-b[1] * data$x) + b[5] * exp(-b[2] * data$x) +
        b[6] * exp(-b[3] * data$x) - data$y
}
lanc.jac <- function(b, data) {
    # 3 exponentials
    expr3 <- exp(-b[1] * data$x)
    expr7 <- exp(-b[2] * data$x)
    expr12 <- exp(-b[3] * data$x)
    J <- matrix(0, nrow = length(data$x), ncol = length(b))
    J[, 4] <- expr3
    J[, 1] <- -(b[4] * (expr3 * data$x))
    J[, 5] <- expr7
    J[, 2] <- -(b[5] * (expr7 * data$x))
    J[, 6] <- expr12
    J[, 3] <- -(b[6] * (expr12 * data$x))
    J
}
bb1 <- c(1, 2, 3, 4, 5, 6)
bb2 <- c(1, 1.1, 1.2, 0.1, 0.1, 0.1)
names(bb1) <- c("a1", "a2", "a3", "c1", "c2", "c3")
names(bb2) <- c("a1", "a2", "a3", "c1", "c2", "c3")
## Check residual function
cat("Sumsquares at bb1 =", as.numeric(crossprod(lanc.res(bb1,
    data = L1))))

## Sumsquares at bb1 = 966.43

require(numDeriv)
JJ <- lanc.jac(bb1, data = L1)
JJn <- jacobian(lanc.res, bb1, data = L1)
cat("max abs deviation JJ and JJn:", max(abs(JJ - JJn)), "\n")

## max abs deviation JJ and JJn: 2.5033e-10

lancexp <- "y ~ (c1*exp(-a1*x) + c2*exp(-a2*x) + c3*exp(-a3*x))"
```

Note that we test the functions. I recommend this be done in EVERY case. In the present example, the tests caught an error in `lanc.res()` where I had the final "+" sign at the beginning of the last line rather than at the end of the penultimate line. A seemingly trivial but important mistake.

Let us now test several different approaches to the problem using both starting vectors.

```
tnls1 <- try(nls(lancexp, start = bb1, data = Lanczos1, trace = FALSE))
strwrap(tnls1)
```

```
## [1] "Error in numericDeriv(form[[3L]], names(ind), env) :"
## [2] "Missing value or an infinity produced when evaluating"
## [3] "the model"
```

```
tnls2 <- try(nls(lancexp, start = bb2, data = Lanczos1, trace = FALSE))
strwrap(tnls2)
```

```
## [1] "Error in nlsModel(formula, mf, start, wts) : singular"
## [2] "gradient matrix at initial parameter estimates"
```

```
## nlmrt
require(nlmrt)
tnlxb1 <- try(nlxb(lancexp, start = bb1, data = Lanczos1, trace = FALSE))
tnlxb1
```

```
## nlmrt class object: x
## residual sumsquares =  5.3038e-09  on  24 observations
##     after  9     Jacobian and  10 function evaluations
##   name        coeff        SE      tstat       pval       gradient     JSingval
## a1         0.887574    0.06331    14.02    3.976e-11    -3.23e-06        4.02
## a2          2.87509    0.05719    50.27           0    -1.765e-05       1.162
## a3          4.96068    0.01666    297.8           0    -1.866e-05      0.2251
## c1        0.0752577    0.00897     8.39    1.236e-07     0.0001546     0.03114
## c2          0.81353    0.01928    42.19           0     9.004e-05    0.002182
## c3           1.6246    0.02808    57.86           0     5.637e-05   0.0001837
```

```
tnlxb2 <- try(nlxb(lancexp, start = bb2, data = Lanczos1, trace = FALSE))
tnlxb2
```

```
## nlmrt class object: x
## residual sumsquares =  4.2906e-06  on  24 observations
##     after  42    Jacobian and  57 function evaluations
##   name        coeff       SE    tstat    pval      gradient     JSingval
## a1         1.87271      NA      NA      NA      4.242e-08        6.007
## a2         1.87274      NA      NA      NA     -3.419e-08         2.57
## a3         4.63965      NA      NA      NA      2.42e-09        0.2743
## c1         4.99146      NA      NA      NA     -1.174e-08       0.02612
## c2        -4.54744      NA      NA      NA     -1.174e-08      1.59e-06
## c3         2.06878      NA      NA      NA     -2.199e-09      1.887e-16
```

```
tnlfb1 <- try(nlfb(bb1, lanc.res, trace = FALSE, data = Lanczos1))
tnlfb1
```

```
## nlmrt class object: x
## residual sumsquares =  5.3036e-09  on  24 observations
##     after  9     Jacobian and  10 function evaluations
##   name        coeff        SE      tstat       pval       gradient     JSingval
## a1         0.887576    0.06331    14.02    3.974e-11    -3.23e-06        4.02
## a2          2.87509    0.05719    50.28           0    -1.765e-05       1.162
## a3          4.96068    0.01666    297.8           0    -1.866e-05      0.2251
## c1        0.0752581    0.00897     8.39    1.235e-07     0.0001546     0.03114
## c2         0.813531    0.01928    42.19           0     9.004e-05    0.002182
## c3           1.6246    0.02808    57.86           0     5.637e-05   0.0001837
```

```
tnlfb2 <- try(nlfb(bb2, lanc.res, trace = FALSE, data = Lanczos1))
tnlfb2

## nlmrt class object: x
## residual sumsquares =  4.2906e-06  on  24 observations
##     after  42    Jacobian and  57 function evaluations
##    name       coeff       SE        tstat       pval       gradient     JSingval
## a1          1.87271     220.7     0.008486     0.9933     4.24e-08      6.007
## a2          1.87274     240.8     0.007776     0.9939    -3.418e-08      2.57
## a3          4.63965    0.06845      67.79          0     2.42e-09      0.2743
## c1          4.99147     346190   1.442e-05          1    -1.174e-08     0.02612
## c2         -4.54745     346190  -1.314e-05          1    -1.174e-08     1.59e-06
## c3          2.06878     0.1006       20.56   5.973e-14    -2.2e-09     9.972e-10

tnlfb1j <- try(nlfb(bb1, lanc.res, lanc.jac, trace = FALSE, data = Lanczos1))
tnlfb1j

## nlmrt class object: x
## residual sumsquares =  5.3038e-09  on  24 observations
##     after  9    Jacobian and  10 function evaluations
##    name       coeff       SE        tstat       pval       gradient     JSingval
## a1          0.887574   0.06331      14.02   3.976e-11    -3.23e-06      4.02
## a2          2.87509    0.05719      50.27          0    -1.765e-05     1.162
## a3          4.96068    0.01666      297.8          0    -1.866e-05     0.2251
## c1          0.0752577  0.00897       8.39   1.236e-07    0.0001546     0.03114
## c2          0.81353    0.01928      42.19          0    9.004e-05     0.002182
## c3          1.6246     0.02808      57.86          0    5.637e-05     0.0001837

tnlfb2j <- try(nlfb(bb2, lanc.res, lanc.jac, trace = FALSE, data = Lanczos1))
tnlfb2j

## nlmrt class object: x
## residual sumsquares =  4.2906e-06  on  24 observations
##     after  42    Jacobian and  57 function evaluations
##    name       coeff       SE      tstat    pval      gradient      JSingval
## a1          1.87271      NA       NA       NA     4.242e-08      6.007
## a2          1.87274      NA       NA       NA    -3.419e-08       2.57
## a3          4.63965      NA       NA       NA     2.42e-09      0.2743
## c1          4.99146      NA       NA       NA    -1.174e-08     0.02612
## c2         -4.54744      NA       NA       NA    -1.174e-08     1.59e-06
## c3          2.06878      NA       NA       NA    -2.199e-09    2.942e-16
```

From the above, we see that it is quite difficult to recover the original exponential parameters (1, 3, and 5). This might be expected, as there is nothing to prevent the three exponential terms from switching roles during the sum-of-squares minimization. Indeed one of the proposed solutions has found what appears to be a saddle point with small gradient, singular Jacobian and two of the exponential parameters are the same. Note that the sum of squares is very small.

nls() should not, of course, be expected to deal with this small-residual problem. However, its "plinear" algorithm does quite well from the start bb1s except for the failure of the relative offset convergence test on the small residuals. The number of iterations has been limited to 20 to reduce output.

```
bb1s <- bb1[1:3] # shorten start, as only 3 starting nonlinear parameters
bb2s <- bb2[1:3]
llexp <- "y~cbind(exp(-a1*x), exp(-a2*x), exp(-a3*x))"
tnls1p <- try(nls(llexp, start = bb1s, data = Lanczos1, algorithm = "plinear",
    trace = TRUE, control = list(maxiter = 20)))
```

```
## 0.007251 :    1.0000  2.0000  3.0000  0.9718 -3.8894   5.3912
## 0.0006056 :   1.29580 -0.04136  4.79435  0.48291 -0.06102  2.10094
## 0.000102 :    2.466172 -0.989412  4.744276  0.667864  0.006105  1.835376
## 1.281e-07 :   2.257394 -0.788374  4.795171  0.620122  0.003522  1.889673
## 9.748e-08 :   2.295432 -0.514802  4.806250  0.633739  0.005439  1.874171
## 8.308e-08 :   2.336692 -0.275700  4.817737  0.647657  0.008083  1.857636
## 7.818e-08 :   2.38021 -0.07137  4.82939  0.66150  0.01150  1.84040
## 7.705e-08 :   2.42499 0.10063 4.84101 0.67502 0.01564 1.82276
## 7.596e-08 :   2.4701 0.2443 4.8525 0.6881 0.0204 1.8049
## 7.317e-08 :   2.51480 0.36385 4.86370 0.70065 0.02561 1.78717
## 6.834e-08 :   2.5584 0.4634 4.8746 0.7127 0.0311 1.7696
## 6.187e-08 :   2.60027 0.54639 4.88513 0.72436 0.03669 1.75239
## 5.444e-08 :   2.64003 0.61569 4.89522 0.73552 0.04221 1.73571
## 5.018e-08 :   2.71463 0.73165 4.91441 0.75636 0.05338 1.70369
## 4.808e-08 :   2.84063 0.88892 4.94806 0.79338 0.07365 1.64642
## 4.261e-09 :   2.99329 1.01211 4.99369 0.85082 0.09646 1.56613
## 2.229e-12 :   3.00003 0.99992 4.99999 0.86072 0.09509 1.55759
## 8.227e-22 :   3.0000 1.0000 5.0000 0.8607 0.0951 1.5576
## 1.429e-25 :   3.0000 1.0000 5.0000 0.8607 0.0951 1.5576
## 1.429e-25 :   3.0000 1.0000 5.0000 0.8607 0.0951 1.5576
## 1.429e-25 :   3.0000 1.0000 5.0000 0.8607 0.0951 1.5576
```

```
strwrap(tnls1p)
```

```
## [1] "Error in nls(llexp, start = bb1s, data = Lanczos1,"
## [2] "algorithm = \"plinear\", : number of iterations"
## [3] "exceeded maximum of 20"
```

```
tnls2p <- try(nls(llexp, start = bb2s, data = Lanczos1, algorithm = "plinear",
    trace = TRUE, control = list(maxiter = 20)))
```

```
## 0.06729 :     1.0   1.1    1.2  271.1 -590.0   321.3
```

```
strwrap(tnls2p)
```

```
## [1] "Error in numericDeriv(form[[3L]], names(ind), env) :"
## [2] "Missing value or an infinity produced when evaluating"
## [3] "the model"
```

The "plinear" algorithm in nls() takes advantage of the linear structure within the model 6.1. In the present case, there are three linear coefficients each multiplying an exponential factor. There is no constant term in the model. We can then think of computing the linear coefficients – the c_j in the model – by linear least squares once we specify the exponential parameters a_j. And there are just three rather than six of these nonlinear parameters.

The linear coefficients c_j are, of course, solutions of a linear least squares problem whose design matrix A has columns $\exp(-a_j * x)$. (Keep in mind that x is a vector of values.) The solution is

$$\mathbf{c} = A^+ * \mathbf{y} \qquad (6.2)$$

where A^+ is the generalized inverse of the design matrix A. The Jacobian, naturally, is constructed from derivatives of the residuals

$$\mathbf{y} - A * \mathbf{c} \qquad (6.3)$$

with respect to the a_j, and this is a quite complicated task. The necessary computational steps have been worked out by Golub and Pereyra (1973), who called the problem *separable nonlinear least squares* and their approach the *variable projection method*. The Fortran implementations have been improved over the years by a number of workers including Fred Krogh, Linda Kaufman, and John Bolstad, but the code is large and unweildy. Recently O'Leary and Rust (2013) presented a new Matlab implementation that markedly improves the readability of the computational steps. It appears that nls() uses a variant of the Golub–Pereyra code. However, it is complicated and in the conventional R distribution is virtually uncommented. This, with the failure of the relative offset convergence test on small residual problems is unfortunate.

My own choice for problems like this is to use numerical approximation of the derivatives with a direct solution of the linear least squares problem. The residual code and example solution is shown in the following.

```
lanclin.res <- function(b, data) {
    # restructured to allow for easier linearization
    xx <- data$x
    yy <- data$y
    res <- rep(NA, length(xx))
    m <- length(xx)
    n <- 3
    A <- matrix(NA, nrow = m, ncol = n)
    for (j in 1:n) {
        A[, j] <- exp(-b[j] * xx)
    }
    lmod <- lsfit(A, yy, intercept = FALSE)
    res <- lmod$residuals
    attr(res, "coef") <- lmod$coef
    res
}
bb1s

## a1 a2 a3
##  1  2  3

res1L <- lanclin.res(bb1s, data = Lanczos1)
cat("sumsquares via lanclin for bb1s start:", as.numeric(crossprod(res1L)),
    "\n")

## sumsquares via lanclin for bb1s start: 0.007251

tnlfbL1 <- try(nlfb(bb1s, lanclin.res, trace = FALSE, data = Lanczos1))
tnlfbL1

## nlmrt class object: x
```

```
## residual sumsquares =  3.2883e-24  on  24 observations
##     after  15    Jacobian and  16 function evaluations
##    name     coeff       SE        tstat       pval      gradient      JSingval
## a1           1    1.219e-09  820097245       0    -2.05e-16      0.03028
## a2           3    1.478e-09  2.029e+09        0    -2.877e-16     0.002831
## a3           5    4.908e-10  1.019e+10        0    -5.729e-17     0.0002005

## Get the linear coefficients
resL1 <- lanclin.res(coef(tnlfbL1), data = Lanczos1)
resL1

##  [1]   3.6607e-13  -4.0313e-13  -4.7523e-13  -2.3124e-15
##  [5]   6.0493e-14   5.7877e-13   5.0700e-13   2.0330e-13
##  [9]   1.4304e-14  -2.4360e-13  -3.8694e-13  -4.6265e-13
## [13]  -4.3367e-13  -2.4682e-13  -1.1390e-13   9.3646e-14
## [17]   2.1907e-13   4.1892e-13   4.6235e-13   4.6641e-13
## [21]   3.2026e-13   8.3785e-14  -2.7687e-13  -7.7064e-13
## attr(,"coef")
##     X1     X2     X3
## 0.0951 0.8607 1.5576

## Now from bb2s
bb2s

## a1  a2  a3
## 1.0 1.1 1.2

res2L <- lanclin.res(bb2s, data = Lanczos1)
cat("sumsquares via lanclin for bb2s start:", as.numeric(crossprod(res2L)),
    "\n")

## sumsquares via lanclin for bb2s start: 0.067291

tnlfbL2 <- try(nlfb(bb2s, lanclin.res, trace = FALSE, data = Lanczos1))
tnlfbL2

## nlmrt class object: x
## residual sumsquares =  2.3014e-24  on  24 observations
##     after  20    Jacobian and  23 function evaluations
##    name     coeff       SE        tstat       pval      gradient      JSingval
## a1           1    1.02e-09   980308128        0    -1.73e-16      0.03028
## a2           3    1.237e-09  2.426e+09        0    -2.295e-16     0.002831
## a3           5    4.106e-10  1.218e+10        0    -6.62e-17      0.0002005

## Get the linear coefficients
resL2 <- lanclin.res(coef(tnlfbL2), data = Lanczos1)
resL2

##  [1]   2.9346e-13  -3.0710e-13  -4.0281e-13  -1.7019e-15
##  [5]   4.7661e-15   5.0202e-13   4.4123e-13   1.6905e-13
##  [9]   1.8683e-14  -2.0428e-13  -3.2395e-13  -3.9082e-13
## [13]  -3.6814e-13  -2.0028e-13  -9.5094e-14   8.0801e-14
## [17]   1.7623e-13   3.5244e-13   3.8359e-13   3.9038e-13
## [21]   2.6474e-13   6.9038e-14  -2.2982e-13  -6.4024e-13
## attr(,"coef")
##     X1     X2     X3
## 0.0951 0.8607 1.5576
```

Did you notice that the sum of squares for start `bb1s` is different from that for `bb1`? This is because we have automatically chosen the best linear parameters for the given nonlinear ones. Note that my approach here only gives variability information (standard errors) for the nonlinear parameters.

Separable least squares problems, or alternatively nonlinear least squares problems where many of the parameters are linear, has a fairly large literature of its own. Some examples and discussion are given in the book by Pereyra and Scherer (2010) or in the article by Mullen et al. (2007).

6.9   Strategies for nonlinear least squares

The discomfort that users face in dealing with nonlinear least squares problems is not that they are generally terribly difficult to solve. Although there are, as we have seen, problems with very wild objective functions, most problems have parameter regions where they behave reasonably well. It is the awkwardness of choice in how to approach such problems that I feel presents the big annoyance.

Using an expression-based tool, the user may give up a lot of flexibility in controlling the loss function while benefiting in many cases from the automatic generation of the Jacobian. Certainly, if the expression is straightforward, then I prefer this line of attack. On the other hand, I find function-based tools, although the initial setup is more work, lend themselves to easier checking and revising, to incorporation into more diverse methods – for instance, by generating the sum of squares and its gradient, we have access to any of the optimization tools.

Bounds constraints seem to be less used than they ought. By putting a box around the solution, we avoid many difficulties, especially if the sum of squares is well defined throughout the region. Masks, that is fixed parameters, exist but appear relatively rarely in R's nonlinear modeling literature.

Variable projection methodology came quite early to R, but as noted is handicapped by the code structure and convergence test in `nls()`. It is my hope to see an updated VARPRO method in an R package in the next couple of years. In the meantime, the choices for modeling when there are many linear parameters are to use `nls()` if the residuals are not near zero or to use numerical approximation of the Jacobian with a function-based method as in the example of the last section.

References

Bates DM and Watts DG 1981 A relative off set orthogonality convergence criterion for nonlinear least squares. Technometrics 23(2), 179–183.

Bates DM and Watts DG 1988 Nonlinear Regression Analysis and Its Applications. John Wiley & Sons, New York.

Golub GH and Pereyra V 1973 The differentiation of pseudo-inverses and nonlinear least squares problems whose variables separate. SIAM Journal of Numerical Analysis 10(2), 413–432.

Huet SS et al. 1996 Statistical Tools for Nonlinear Regression: A Practical Guide with S-PLUS Examples, Springer series in statistics. Springer-Verlag, Berlin & New York.

Lanczos C 1956 Applied Analysis. Prentice-Hall, Englewood Cliffs, NJ. Reprinted by Dover, New York, 1988.

Mullen KM, Vengris M and Stokkum IH 2007 Algorithms for separable nonlinear least squares with application to modelling time-resolved spectra. Journal of Global Optimization 38(2), 201–213.

Nash JC 1979 Compact Numerical Methods for Computers: Linear Algebra and Function Minimisation. Adam Hilger, Bristol. Second Edition, 1990, Bristol: Institute of Physics Publications.

Nash JC 2012 Letter: weigh less in tonnes? Significance 9(6), 45.

Nash JC and Price K 1979 Fitting two straight lines. In Proceedings of Computer Science and Statistics: 12th annual Symposium on the Interface: May 10 & 11, 1979 (ed. J. Gentleman) University of Waterloo, Waterloo, Ontario, Canada, pp. 363–367. University of Waterloo, Ontario, Canada. JNFile: 78Fitting2StraightLines.pdf. Page numbers may be in error. Also Engineering and Statistical Research Institute, Agriculture Canada, Contribution No. I-99, 1978.

Nash JC and Walker-Smith M 1987 Nonlinear Parameter Estimation: An Integrated System in BASIC. Marcel Dekker, New York. See http://www.nashinfo.com/nlpe.htm for an expanded downloadable version.

O'Leary DP and Rust BW 2013 Variable projection for nonlinear least squares problems. Computational Optimization and Applications 54(3), 579–593.

Pereyra V and Scherer G 2010 Exponential Data Fitting and its Applications. Bentham Science Publishers, Sharjah.

Ratkowsky DA 1983 Nonlinear Regression Modeling: A Unified Practical Approach. Marcel Dekker Inc., New York and Basel.

Ritz C and Streibig JC 2008 Nonlinear Regression with R. Springer, New York.

Ross GJ 1990 Nonlinear Estimation. Springer-Verlag, New York.

Seber GAF and Wild CJ 1989 Nonlinear Regression. John Wiley & Sons, New York.

Venables WN and Ripley BD 1994 Modern Applied Statistics with S-PLUS. Springer-Verlag, Berlin, Germany / London.

7

Nonlinear equations

Nonlinear equations (NLEs) in more than one unknown parameter are the subject of this chapter. This is the multiparameter version of root-finding from Chapter 4. Generally, I have NOT found this to be, of itself, a prominent problem for statistical workers, but there are some important uses. Unfortunately, people sometimes try to use NLEs methods to find extrema of nonlinear functions. For such problems, my experience suggests that it is almost always better to use an optimization tool.

As we have already dealt with one-parameter problems in Chapter 4, we will be dealing with two or more equations in an equal number of unknowns. If we consider that these functions are like **residuals** in the nonlinear least squares problem and we write these equations in the n parameters $\mathbf{x}$ as

$$\mathbf{r}(\mathbf{x}) = \mathbf{0} \qquad (7.1)$$

then clearly a least squares solution that has a zero sum of squares is also a solution to the NLEs problem, so nonlinear least squares methods can be considered for NLEs problems, but we must be careful to check that a solution has indeed been found, and there may be efficiencies in using methods that are explicitly intended for the NLE problem, as a least squares approach in some sense squares quantities that should be considered in their natural scale. However, my opinion is that for many problems, the nonlinear least squares approach to NLEs problems is useful as a natural check on solutions and a measure of how "bad" proposed solutions may be. It does not, of course, give any direct help in dealing with multiple solutions.

7.1 Packages and methods for nonlinear equations

There are two R packages that I have used that are explicitly for solving NLEs, BB and `nleqslv`.

Nonlinear Parameter Optimization Using R Tools, First Edition. John C. Nash.
© 2014 John Wiley & Sons, Ltd. Published 2014 by John Wiley & Sons, Ltd.
Companion Website: www.wiley.com/go/nonlinear_parameter

7.1.1 `BB`

This package (Varadhan and Gilbert, 2009a) is based on a special gradient method of Cruz et al. (2006), which has proven effective for a number of large-scale NLE systems and exploits a novel stepsize choice introduced by Barzilai and Borwein (1988). Within the package are two methods for NLEs, namely, the function `sane()` (spectral approach to nonlinear equations) that requires a directional derivative of the residual sum of squares and `dfsane()` that is derivative free. In my experience, both functions work quite well. The package also includes a function minimization routine, `spg`, based on the same ideas.

7.1.2 `nleqslv`

This package (Hasselman, 2013) is built on several tools associated with Dennis and Schnabel (1983). It uses ideas from the quasi-Newton family of methods applied directly to the NLEs, whereas for optimization, we aim to solve the equations that set the gradient to the null vector. `nleqslv()` can use analytic Jacobian information if supplied by the user. There are various controls. In particular, the method can be "Broyden" or "Newton." The former starts with an approximate Jacobian and tries to update or improve this approximation at each iteration in the manner of quasi-Newton methods. The "Newton" choice calculates a Jacobian at each iteration. There are also control choices for the line search/trust region strategy. While all these choices are useful to the expert, they provide much to confuse the unfamiliar user.

Neither `BB` nor `nleqslv` offers bounds constraints to prevent the methods from finding inadmissible solutions or wandering into unacceptable parameter regions.

7.1.3 Using nonlinear least squares

As mentioned earlier, we can use nonlinear least squares methods to solve NLEs. Over the years, I have preferred to do this but must acknowledge that the preference may be driven by my familiarity with such methods. It is, of course, necessary to check that the sum of squares is essentially zero at the proposed solution. Moreover, some nonlinear least squares tools and in particular R's `nls()` function are NOT designed to deal with small-residual problems.

Note that tools such as those in package `nlmrt` can deal with bounds on the parameters, which may be helpful in finding admissible solutions or excluding inadmissible ones.

7.1.4 Using function minimization methods

One further level higher is the application of general function minimization tools to the sum-of-squares function of the equations. Here we are ignoring the special structure and must once more check that the objective is zero at a proposed solution. While not the ideal approach, general tools may be useful in allowing constraints to be applied when these are important.

7.2   A simple example to compare approaches

(Dennis and Schnabel 1983, Example 6.5.1, p. 149) present a simple example of two equations that we can use to illustrate the above suggestions. We write the equations as residuals and will then try to find solutions that have $r = 0$.

$$r_1 = x_1^2 + x_2^2 - 2$$
$$r_2 = \exp(x_1 - 1) + x_2^3 - 2 \qquad (7.2)$$

We first set up the residuals and their partial derivatives (Jacobian).

```
# Dennis Schnabel example 6.5.1 page 149
dslnex <- function(x) {
    r <- numeric(2)
    r[1] <- x[1]^2 + x[2]^2 - 2
    r[2] <- exp(x[1] - 1) + x[2]^3 - 2
    r
}
jacdsln <- function(x) {
    n <- length(x)
    Df <- matrix(numeric(n * n), n, n)
    Df[1, 1] <- 2 * x[1]
    Df[1, 2] <- 2 * x[2]
    Df[2, 1] <- exp(x[1] - 1)
    Df[2, 2] <- 3 * x[2]^2
    Df
}
ssdsln <- function(x) {
    ## a sum of squares function for 6.5.1 example
    rr <- dslnex(x)
    val <- as.double(crossprod(rr))
}
```

Now we will try to solve these with the NLE tools from BB and nleqslv.

```
require(microbenchmark) # for timing
require(BB)
require(nleqslv)
xstart <- c(2, 0.5)
fstart <- dslnex(xstart)
## xstart:
print(xstart)

## [1] 2.0 0.5

## resids at start:
print(fstart)

## [1] 2.25000 0.84328

# a solution is c(1,1)
print(dslnex(c(1, 1)))

## [1] 0 0
```

We will now attempt to find this solution by various methods. While doing so, we shall use `microbenchmark` to get the mean and standard deviation of the execution time of each method. We include the syntax of the calls but suppress display of the data manipulation to save space and clutter.

First, let us try the "default" solver of package BB.

```
tabbx <- microbenchmark(abbx <- BBsolve(xstart, dslnex, quiet = TRUE))$time
cat("solution at ", abbx$par[1], " ", abbx$par[2], "\n")

## solution at  1    1
```

But BB can use its functions `sane()` and `dfsame()` directly. Note the `quiet` and `trace` settings to avoid unnecessary output here.

```
cl <- list(trace = FALSE)
# sane
tasx <- microbenchmark(asx <- sane(xstart, dslnex, quiet = TRUE,
    control = cl))$time
cat("solution at ", asx$par[1], " ", asx$par[2], "\n")

## solution at  1    1

tadx <- microbenchmark(adx <- dfsane(xstart, dslnex, quiet = TRUE,
    control = cl))$time
cat("solution at ", adx$par[1], " ", adx$par[2], "\n")

## solution at  1    1
```

Another option is package `nleqslv`. There is a default method, but there are many options, of which we try the two `method` choices "Newton" and "Broyden."

```
require(nleqslv)
tanlx <- microbenchmark(anlx <- nleqslv(xstart, dslnex))$time
cat("solution at ", anlx$x[1], " ", anlx$x[2], "\n")

## solution at  1    1

tanlNx <- microbenchmark(anlNx <- nleqslv(xstart, dslnex,
 method = "Newton"))$time
cat("solution at ", anlNx$x[1], " ", anlNx$x[2], "\n")

## solution at  1    1

tanlBx <- microbenchmark(anlBx <- nleqslv(xstart, dslnex,
 method = "Broyden"))$time
cat("solution at ", anlBx$x[1], " ", anlBx$x[2], "\n")

## solution at  1    1
```

As we have noted, a least squares solution at zero is clearly a solution of the NLEs. Packages `minpack.lm` and `nlmrt` offer functions `nls.lm` and `nlfb` to deal with problems specified as residual functions written in R. There is also a function `gaussNewton()` in `pracma` that I have not yet tried. There are other functions to deal with problems specified via expressions. In the following trials, we see that the nonlinear least squares programs sometimes find "answers" that do not solve the nonlinear equations.

```
require(minpack.lm)
takx <- microbenchmark(akx <- nls.lm(par = xstart, fn = dslnex,
    jac = jacdsln))$time
cat("solution at ", akx$par[1], " ", akx$par[2],
"    with equation residuals \n")

## solution at  1.4851   2.5784e-06    with equation residuals

print(dslnex(akx$par))

## [1]  0.20548 -0.37569


taknx <- microbenchmark(aknx <- nls.lm(par = xstart, fn = dslnex,
    jac = NULL))$time
cat("solution at ", aknx$par[1], " ", aknx$par[2],
"    with equation residuals \n")

## solution at  1.4851   4.5665e-05    with equation residuals

print(dslnex(aknx$par))

## [1]  0.20546 -0.37570


require(nlmrt)
tajnx <- microbenchmark(ajnx <- nlfb(start = xstart, dslnex,
    jacdsln))$time
ajnxc <- coef(ajnx)
cat("solution at ", ajnxc[1], " ", ajnxc[2], "  with equation residuals \n")

## solution at  1   1    with equation residuals

print(dslnex(ajnx$coefficients))

## [1]   9.5312e-10 -1.0427e-09


tajnnx <- microbenchmark(ajnnx <- nlfb(start = xstart, dslnex,
    jac = NULL))$time
ajnnxc <- coef(ajnnx)
cat("solution at ", ajnnxc[1], " ", ajnnxc[2],
"    with equation residuals \n")

## solution at  1   1    with equation residuals

print(dslnex(ajnnx$coefficients))

## [1]   9.5315e-10 -1.0427e-09
```

A final layer of possible solution methods is to simply minimize the Sum-of-squares function using a general function minimization tool. Here we use all the methods in `optimx` in brute force manner. However, we do instruct the package to use the `numDeriv` approximation to derivatives.

```
require(optimx)

## Loading required package: optimx

aox <- suppressWarnings(optimx(xstart, ssdsln, gr = "grcentral",
    method = "all"))

## Loading required package: numDeriv

summary(aox, order = value)

##                    p1          p2  value fevals gevals niter
## hjkb            1.000  1.000e+00  0.0000    120     NA     19
## newuoa          1.485  2.021e-08  0.1834     52     NA     NA
## bobyqa          1.485 -1.089e-06  0.1834     63     NA     NA
## nlm             1.485 -4.973e-07  0.1834     NA     NA     19
## BFGS            1.485  1.163e-05  0.1834     24     24     NA
## CG              1.485  1.163e-05  0.1834     24     24     NA
## Nelder-Mead     1.485  1.163e-05  0.1834     24     24     NA
## L-BFGS-B        1.485  1.163e-05  0.1834     24     24     NA
## Rcgmin          1.485 -2.285e-05  0.1834     47     24     NA
## Rvmmin          1.485 -2.771e-05  0.1834     33     15     NA
## ucminf          1.485  3.486e-05  0.1834     14     14     NA
## nlminb          1.485 -3.600e-05  0.1834     13     10     10
## spg             1.485 -8.065e-05  0.1834     39     NA     37
## nmkb            1.485  2.765e-06  0.1834     55     NA     NA
##                 convcode  kkt1   kkt2 xtimes
## hjkb                   0  TRUE   TRUE  0.004
## newuoa                 0  TRUE   TRUE  0.000
## bobyqa                 0  TRUE  FALSE  0.004
## nlm                    0  TRUE  FALSE  0.004
## BFGS                   0  TRUE  FALSE  0.004
## CG                     0  TRUE  FALSE  0.004
## Nelder-Mead            0  TRUE  FALSE  0.004
## L-BFGS-B               0  TRUE  FALSE  0.000
## Rcgmin                 0  TRUE   TRUE  0.004
## Rvmmin                 0  TRUE  FALSE  0.004
## ucminf                 0  TRUE  FALSE  0.000
## nlminb                 0  TRUE  FALSE  0.004
## spg                    0  TRUE  FALSE  0.056
## nmkb                   0 FALSE   TRUE  0.004
```

What can we say about all the optimization solutions? First, most methods have NOT found a solution to the NLEs. We also suspect that the solution by `hjkb()` of package `dfoptim` is lucky. However, `optimx` has given us the KKT tests that provide a measure of reassurance that answers are at least local minima of the sum of squares.

Similarly, `nls.lm`, both with analytic and numerically approximated Jacobian, has found an answer that is NOT a solution of the NLEs. However, a little probing of the solution shows

```
JJ <- jacdsln(akx$par)
svd(JJ)$d
```

```
## [1] 3.3853e+00 2.4743e-06
```

```
res <- dslnex(akx$par)
res
```

```
## [1]  0.20548 -0.37569
```

```
g <- t(JJ) %*% res
g
```

```
##             [,1]
## [1,] 6.8001e-05
## [2,] 1.0596e-06
```

```
fn <- ssdsln(akx$par)
fn
```

```
## [1] 0.18336
```

```
HH <- hessian(ssdsln, akx$par)
HH
```

```
##             [,1]        [,2]
## [1,] 2.2522e+01 3.0682e-05
## [2,] 3.0682e-05 8.2190e-01
```

```
eigen(HH)$values
```

```
## [1] 22.5220  0.8219
```

Thus the gradient of the sum of squares at the reported answer is nearly null, and the Hessian is positive definite. We have found a local minimum of the sum-of-squares function, but it is not a solution of the NLEs. Again, for the moment, we will not search for the reasons why nls.lm has converged to the particular point returned.

Which of the methods is the most efficient? We have used microbenchmark, so we have the timing on a particular machine (in Chapter 18, we see how to calibrate such timings to that machine). Note that the output from optimx above includes a single timing for each method, while microbenchmark provides a set of such timings. As the optimization results were generally unacceptable, we will not deal with them further.

We also have some data on the number of residual evaluations (fevals), Jacobian evaluations (jevals), or "iterations." Unfortunately, the last element is not comparable between methods. We also may have some return code from a method that may provide information on the validity of the returned solution. Note that the output from optimx above includes similar information.

First, let us look at the returned parameters and residuals. This restates what we already know, that is, that all but nls.lm and optimx return the solution at c(1, 1).

```
##                par1    par2       res1       res2
## BBsolve      1.0000 1.00e+00 0.0000001 -0.0000001
## sane         1.0000 1.00e+00 0.0000000  0.0000000
## dfsane       1.0000 1.00e+00 0.0000000  0.0000000
## nleqslv      1.0000 1.00e+00 0.0000000  0.0000000
## nleqslvN     1.0000 1.00e+00 0.0000000  0.0000000
## nleqslvB     1.0000 1.00e+00 0.0000000  0.0000000
## nls.lm       1.4851 2.60e-06 0.2054770 -0.3756872
## nls.lm-noJ   1.4851 4.57e-05 0.2054591 -0.3756970
## nlfb         1.0000 1.00e+00 0.0000000  0.0000000
## nlfb-noJ     1.0000 1.00e+00 0.0000000  0.0000000
```

The timing for the different methods – mean and standard deviation in milliseconds – over the default 100 times each method is run by `microbench-mark` is as follows.

```
##             mean(t)  sd(t)
## BBsolve        2343    300
## sane           1573    274
## dfsane          974     47
## nleqslv        1076   1925
## nleqslvN        451     25
## nleqslvB        502     42
## nls.lm          733     32
## nls.lm-noJ      777     35
## nlfb           5834    573
## nlfb-noJ       6252    344
```

This timing data largely reflects the nature and implementation of the methods. `nlfb` and `nls.lm` are very similar methods, but the latter is implemented in `Fortran` and the former is implemented in R and designed to aggressively seek solutions. Both are methods for nonlinear least squares. `nleqslv` is intended to solve NLEs and it is implemented in Fortran. If there are any surprises, it is that

- the standard deviations of times are quite large, suggesting that it is difficult to time methods accurately in a general-purpose computer,

- none of the approaches takes very much time. These are fairly inexpensive tasks.

The count and status data are

```
##             fevals jevals niter ccode
## BBsolve         68     NA     6     0
## sane            33     NA    16     0
## dfsane          17     NA    16     0
## nleqslv         12      1    10     1
## nleqslvN         6      5     5     1
## nleqslvB        12      1    10     1
## nls.lm          NA     NA    11    NA
## nls.lm-noJ      NA     NA    12    NA
## nlfb            11      7    NA    NA
## nlfb-noJ        11      7    NA    NA
```

These are, as noted, not strictly comparable, but the level of effort for all the methods is of the same order of magnitude. To make a careful comparison, we would need to insert code in the residual and Jacobian codes to record the actual calls to those routines.

7.3   A statistical example

Varadhan and Gilbert (2009b) present an example (using simulated data) of Poisson regression with offset, where we want to solve the estimating equations

$$\sum_{i=1}^{n} \{X_i^T \{Y_i - t_i * \exp(X_i^T \beta)\}\} = 0 \tag{7.3}$$

These are the equations for the maximizing the likelihood of the counts Y_i at observation i, which are observed at times t_i where the covariates are X_i. The β are our parameters that we wish to determine. Because the Y_i are counts, we cannot expect the residuals

$$Y_i - t_i * \exp(X_i^T \beta) \tag{7.4}$$

to be zero. There is an "offset" defined by the times t_i. The following code runs the solution via dfsane() from BB as well as giving the solution via the generalized linear models package glm and the minimization of the appropriate negative log likelihood. Thus there are several acceptable ways to solve these problems. Details can, and very often do, create ample opportunities for "silly" errors. It is often a good idea to try multiple solution methods, at least on some trial problems.

```
# Poisson estimating equation with 'offset'
U.eqn <- function(beta, Y, X, obs.period) {
    Xb <- c(X %*% beta)
    as.vector(crossprod(X, Y - (obs.period * exp(Xb)))) # changed 130129
}
poisson.sim <- function(beta, X, obs.period) {
    Xb <- c(X %*% beta)
    mean <- exp(Xb) * obs.period
    rpois(nrow(X), lambda = mean)
}
require(BB)
require(setRNG)
# this RNG setting can be used to reproduce results
test.rng <- list(kind = "Mersenne-Twister", normal.kind = "Inversion",
    seed = 1234)
old.seed <- setRNG(test.rng)
n <- 500
X <- matrix(NA, n, 8)
X[, 1] <- rep(1, n)
X[, 3] <- rbinom(n, 1, prob = 0.5)
X[, 5] <- rbinom(n, 1, prob = 0.4)
X[, 7] <- rbinom(n, 1, prob = 0.4)
X[, 8] <- rbinom(n, 1, prob = 0.2)
X[, 2] <- rexp(n, rate = 1/10)
```

```
X[, 4] <- rexp(n, rate = 1/10)
X[, 6] <- rnorm(n, mean = 10, sd = 2)
obs.p <- rnorm(n, mean = 100, sd = 30) # observation period
beta <- c(-5, 0.04, 0.3, 0.05, 0.3, -0.005, 0.1, -0.4)
Y <- poisson.sim(beta, X, obs.p)
## Using dfsane from BB
aBB <- dfsane(par = rep(0, 8), fn = U.eqn, control = list(NM = TRUE,
    M = 100, trace = FALSE), Y = Y, X = X, obs.period = obs.p)
aBB


## $par
## [1] -5.0157227  0.0424482  0.3082518  0.0492517  0.3184578
## [6] -0.0055035  0.0747347 -0.4613516
##
## $residual
## [1] 2.4552e-08
##
## $fn.reduction
## [1] 9141
##
## $feval
## [1] 1748
##
## $iter
## [1] 1377
##
## $convergence
## [1] 0
##
## $message
## [1] "Successful convergence"

# 'glm' gives same results as solving the estimating
# equations
ans.glm <- glm(Y ~ X[, -1], offset = log(obs.p), family =
poisson(link = "log"))
ans.glm

##
## Call:  glm(formula = Y ~ X[, -1], family = poisson(link = "log"),
## offset = log(obs.p))
##
## Coefficients:
## (Intercept)      X[, -1]1      X[, -1]2      X[, -1]3
##     -5.0157        0.0424        0.3083        0.0493
##       X[, -1]4      X[, -1]5      X[, -1]6      X[, -1]7
##        0.3185       -0.0055        0.0747       -0.4614
##
## Degrees of Freedom: 499 Total (i.e. Null);   492 Residual
## Null Deviance:      2170
## Residual Deviance: 519   AIC: 1660
```

As before, nonlinear least squares and general optimization offer other approaches to a solution by seeking a minimal sum of squares of the residuals (constraint violations) that is zero.

```
## nonlinear least squares on residuals
ssfn <- function(beta, Y, X, obs.p) {
    resfn <- U.eqn(beta, Y, X, obs.p)
    as.double(crossprod(resfn))
}
ansopt <- optim(beta, ssfn, control = list(trace = 0, maxit = 5000),
    Y = Y, X = X, obs.p = obs.p)
ansopt

## $par
## [1] -5.0263739  0.0422083  0.2379272  0.0495724  0.3418682
## [6] -0.0011892  0.0755324 -0.5124484
##
## $value
## [1] 682.68
##
## $counts
## function gradient
##     3009       NA
##
## $convergence
## [1] 0
##
## $message
## NULL


negll <- function(beta, Y = Y, X = X, obs.p = obs.p) {
    # log likelihood
    Xb <- c(X %*% beta)
    ll <- crossprod(Y, Xb) - sum(obs.p * exp(Xb))
    nll <- -as.double(ll)
}
require(optimx)
print(negll(beta, Y, X, obs.p = obs.p))

## [1] 5987.9


ansopll <- optimx(beta, negll, gr = NULL, control = list(trace = 0,
    usenumDeriv = TRUE, all.methods = TRUE), Y = Y, X = X, obs.p = obs.p)

## Warning: Replacing NULL gr with 'numDeriv' approximation
## Warning: bounds can only be used with method L-BFGS-B (or Brent)
## Warning: bounds can only be used with method L-BFGS-B (or Brent)
## Warning: bounds can only be used with method L-BFGS-B (or Brent)
## Warning: NA/Inf replaced by maximum positive value
## Warning: NA/Inf replaced by maximum positive value
## Warning: NA/Inf replaced by maximum positive value
## Warning: Rcgmin - undefined function
## Warning: Rcgmin - undefined function
## Warning: Rcgmin - undefined function
## Warning: Too many gradient evaluations

## Note: trace=1 gives TEX error 'dimension too large'.
ansopll
```

```
##                  p1       p2       p3        p4       p5
## BFGS        -5.0102 0.042455 0.30908 0.049251 0.31457
## CG          -5.0102 0.042455 0.30908 0.049251 0.31457
## Nelder-Mead -5.0102 0.042455 0.30908 0.049251 0.31457
## L-BFGS-B    -5.0102 0.042455 0.30908 0.049251 0.31457
## nlm         -5.0157 0.042447 0.30826 0.049251 0.31847
## nlminb      -5.0157 0.042448 0.30825 0.049252 0.31846
## spg              NA       NA      NA       NA       NA
## ucminf      -5.0157 0.042448 0.30825 0.049252 0.31846
## Rcgmin      -5.0157 0.042448 0.30825 0.049252 0.31846
## Rvmmin      -5.0157 0.042448 0.30825 0.049252 0.31846
## newuoa      -5.0156 0.042448 0.30824 0.049252 0.31843
## bobyqa      -5.0158 0.042448 0.30823 0.049252 0.31849
## nmkb        -5.0157 0.042448 0.30828 0.049251 0.31848
## hjkb        -5.0155 0.042449 0.30826 0.049252 0.31845
##                  p6        p7       p8      value fevals
## BFGS        -0.0057858 0.071578 -0.46005 5.9859e+03     46
## CG          -0.0057858 0.071578 -0.46005 5.9859e+03     46
## Nelder-Mead -0.0057858 0.071578 -0.46005 5.9859e+03     46
## L-BFGS-B    -0.0057858 0.071578 -0.46005 5.9859e+03     46
## nlm         -0.0055071 0.074737 -0.46134 5.9859e+03     NA
## nlminb      -0.0055035 0.074734 -0.46135 5.9859e+03     74
## spg                NA       NA       NA 8.9885e+307     NA
## ucminf      -0.0055035 0.074735 -0.46135 5.9859e+03     25
## Rcgmin      -0.0055044 0.074734 -0.46135 5.9859e+03   1501
## Rvmmin      -0.0055036 0.074735 -0.46135 5.9859e+03     83
## newuoa      -0.0055131 0.074731 -0.46140 5.9859e+03   3031
## bobyqa      -0.0054958 0.074761 -0.46135 5.9859e+03   3607
## nmkb        -0.0055071 0.074735 -0.46133 5.9859e+03   1532
## hjkb        -0.0055264 0.074731 -0.46136 5.9859e+03   2660
##             gevals niter convcode kkt1 kkt2 xtimes
## BFGS            46    NA        0 TRUE TRUE  0.396
## CG              46    NA        0 TRUE TRUE  0.280
## Nelder-Mead     46    NA        0 TRUE TRUE  0.264
## L-BFGS-B        46    NA        0 TRUE TRUE  0.312
## nlm             NA    67        0 TRUE TRUE  0.156
## nlminb          51    50        0 TRUE TRUE  0.444
## spg             NA    NA     9999   NA   NA  0.152
## ucminf          25    NA        0 TRUE TRUE  0.144
## Rcgmin         319    NA        1 TRUE TRUE  3.941
## Rvmmin         501    NA        1 TRUE TRUE  4.780
## newuoa          NA    NA        0 TRUE TRUE  1.124
## bobyqa          NA    NA        0 TRUE TRUE  2.016
## nmkb            NA    NA        0 TRUE TRUE  1.336
## hjkb            NA    19        0 TRUE TRUE  1.080
```

References

Barzilai J and Borwein JM 1988 Two-point step size gradient methods. IMA Journal of Numerical Analysis 8(1), 141–148.

Cruz WL, Martínez JM and Raydan M 2006 Spectral residual method without gradient information for solving large-scale nonlinear systems: theory and experiments. Mathematics of Computation 75, 1429–1448.

Dennis JE and Schnabel RB 1983 Numerical Methods for Unconstrained Optimization and Nonlinear Equations. Prentice-Hall, Englewood Cliffs, NJ.

Hasselman B 2013 *nleqslv: Solve systems of non linear equations*. R package version 2.0.

Varadhan R and Gilbert P 2009a BB: an R package for solving a large system of nonlinear equations and for optimizing a high-dimensional nonlinear objective function. Journal of Statistical Software 32(4), 1–26.

Varadhan R and Gilbert P 2009b BB: an R package for solving a large system of nonlinear equations and for optimizing a high-dimensional nonlinear objective function. Journal of Statistical Software 32(4), 1–26.

8

Function minimization tools in the base R system

The scientists who built the core R system have included a number of tools for optimization. In this chapter, we explore these and review their strengths and weaknesses. It is, however, important to note the following.

- All the methods discussed in this chapter are relatively "old." They date from the 1970s to 1980s at the latest, even if the codes are somewhat more recent. This does not mean that they are not good tools. It is rather that there are more recent tools that take account of changing requirements, deal with larger or somewhat different problems, and include capabilities that the base tools lack.

- The base methods have been used by other R tools, so are present in some cases as support for those tools.

- It is very difficult to remove something from a working system, even if there are more suitable replacements.

8.1 `optim()`

`optim()` is one of the most used tools in R. It incorporates five methods, three of which are derived from the author's own work (Nash, 1979). The five methods are largely independent but are all called through `optim()`, using a call such as

```
ans<-optim(start, fn=myfn, gr=mygr, method="BFGS",
        X=myx, Y=myY, control=list(trace=1))
```

Nonlinear Parameter Optimization Using R Tools, First Edition. John C. Nash.
© 2014 John Wiley & Sons, Ltd. Published 2014 by John Wiley & Sons, Ltd.
Companion Website: www.wiley.com/go/nonlinear_parameter

where start is the vector of parameters that are used as the initial point of the optimization iteration, myfn is the function to be minimized, mygr is its gradient, and the control list specifies that we want to follow the iterations (trace), while the elements given by X=myX, Y=myY are named variables required by myfn and mygr and are passed to optim() via the "..." arguments. The method is specified via a character string, here using a method called "BFGS" (Broyden–Fletcher–Goldfarb–Shanno).

The methods included in optim() are

- *Nelder–Mead.* This is the default method and is a variant of that described in Nelder and Mead (1965). This method requires no derivative information to be supplied; indeed it will be ignored. The method builds a polytope (or simplex) of $(n + 1)$ points in n-space, then reflects the "highest" point through the centroid of the rest of the points, and proceeds to use such geometric information to steadily reduce the function value. The name "Nelder–Mead Simplex method" has in the past caused confusion with the linear programming method of Dantzig (Nash, 2000). I prefer to use "Nelder–Mead polytope method" or simply "Nelder–Mead" or even "NM."

- *BFGS.* This is a variable metric method due to Fletcher (1970). This method is a descent method using the gradient, and analytic gradients or very good approximations are recommended. Starting with a simple identity matrix, gradient information built up during the iterations of this method is used to build an approximate inverse Hessian, so that the method "approaches" Newton's method. While experiments show the approximation to be poor, in practice the method performs well. The use of a BFGS formula to "update" the inverse Hessian approximation has led to the naming of this method in R, but readers will more likely find useful information about this method by using the search term "variable metric."

- *L-BFGS-B.* This method is the most sophisticated in the optim() set. It uses a limited-memory update of the inverse Hessian so is somewhat similar to the BFGS approach. However, the amount of information kept to build the update can be controlled by the user, the default being to keep information from the last five iterations. See Lu et al. (1994) or the code published by Zhu et al. (1997). The method can, moreover, deal with bounds on the parameters and is the default method if any bound is specified. Generally, L-BFGS-B works well, although its very sophistication does get in the way of success on occasions. At the time of writing, R does not include the updated version of this method (Morales and Nocedal, 2011).

- *CG.* This is a conjugate gradient descent method. It is based on various codes extant in the 1960s and 1970s. As with BFGS, we start with the gradient as a search direction and then modify this at each iteration. Where the BFGS/variable metric method updated a full matrix, CG simply builds the new search direction from the old one and new and old gradient information.

This makes it very attractive for problems in large numbers of parameters. There was much discussion about which of the three main candidates can be used for the computation of the new search direction in papers of the era. These are the Fletcher–Reeves, Polak–Ribière, and Beale–Sorenson updates. (Note that other names are also used, just to confuse us all.) R allows a choice of update via a control-list item type which takes values 1 (default), 2, or 3 respectively. Personally, I have found the second option preferable, but none are definitely better than the others. Overall I have been disappointed in this method. However, a more recent approach (Dai and Yuan, 2001) that cleverly combines and exploits differences in the updates allowed me to prepare **Rcgmin** with which I have been more than pleased.

- *SANN*. This is a stochastic method that I believe does NOT belong with optim(). First, it will always evaluate the function a fixed number of times and then report "convergence." This is exceedingly misleading to most users. Moreover, the idea of "temperature" in the process is inconsistent with the descent ideas of the other methods.

8.2 nlm()

nlm is an interface to a particularized version of UNCMIN, a set of algorithms for unconstrained minimization from Schnabel et al. (1985). The approach is a Newton or quasi-Newton method, but the details are somewhat difficult to understand from the manual. The original code provided for a variety of line search and updating techniques, as well as different gradient approximations.

The calling sequence for nlm() is rather different from that of optim(), and the manual gives the usage as

```
nlm(f, p, ..., hessian = FALSE, typsize = rep(1, length(p)),
    fscale = 1, print.level = 0, ndigit = 12, gradtol = 1e-6,
    stepmax = max(1000 * sqrt(sum((p/typsize)^2)), 1000),
    steptol = 1e-6, iterlim = 100, check.analyticals = TRUE)
```

Note that the function f is given first in the sequence, and the parameters second, a reverse from optim(). Moreover, the gradient and possibly Hessian code are passed as attributes to f, described as

```
f: the function to be minimized.  If the function value has an
   attribute called 'gradient' or both 'gradient' and 'hessian'
   attributes, these will be used in the calculation of updated
   parameter values.  Otherwise, numerical derivatives are used.
   'deriv' returns a function with suitable 'gradient'
   attribute.  This should be a function of a vector of the
   length of 'p' followed by any other arguments specified by
   the '...' argument.
```

As with so many optimization routines, nlm() has a lot of controls that essentially adjust heuristics aimed at tuning the methods. In practice, I have rarely if ever used these, even though I work in the optimization field and should have the expertise to do so. The reality is that each program has different names or ways of implementing relatively similar ideas that concern how "big" or "small" different quantities should be. Users find it tedious to learn the intricacies of each control, and tinkering is time consuming and generally unproductive.

A possible exception is the limit on iterations. I generally want to know if a method gets to a valid solution when allowed to take as many iterations as it likes.

Despite the above criticisms of `nlm()`, in my experience, it is one of the more reliable and efficient optimization methods for unconstrained functions, even when only the function code is provided. To simplify its use, I generally call it via `optimx()`, which allows the use of a common, `optim`-like calling sequence.

8.3   `nlminb()`

Because `nlm()` is an unconstrained minimizer, but bounds constraints were needed for a number of maximum likelihood problems, `nlminb()` was added to the base set of optimizers. The code on which `nlminb()` is based was written by David Gay at Bell Labs and was part of the PORT Fortran Library (Fox et al. 1978). The S language at Bell Labs used this as its optimizer, and Doug Bates, in an e-mail dated June 2008, states that

```
> I created the nlminb function because I was unable to get reliable
> convergence on some difficult optimization problems for the nlme and
> lme4 packages using nlm and optim. The nlme package was originally
> written for S from Bell Labs (the forerunner of SPLUS) and the PORT
> package was the optimization code used. Even though it is very old
> style Fortran code I find it quite reliable as an optimizer.
```

I agree with this general view, although the code in the PORT library is not easy to follow and use. Moreover, as with `nlm()`, there are many settings and choices that could be helpful, but more R users will want to leave alone. However, `nlminb` is quite similar in call to `optim()`, namely,

```
nlminb(start, objective, gradient = NULL, hessian = NULL, ...,
   scale = 1, control = list(), lower = -Inf, upper = Inf)
```

The issue of controls was highlighted in a discovery in 2010 (see https://stat.ethz.ch/pipermail/r-help/2010-July/245194.html) that `nlminb()` assumed positive objective functions (likely sums of squares) and that a default test for termination was $|f(x)| <$ `abs.tol`, a control parameter to `nlminb`. Many negative log likelihood functions will not be positive, and it is difficult to know how many incorrect results were used or published. The issue underlines one that is central to open-source software such as R – we need to be able to easily review code if it is to be improved. The PORT library has many fine codes, but in my view, they are not easily understood or changed, especially as implemented for the use by R, where the interface adds another layer of confusion.

8.4   Using the base optimization tools

Let us solve the well-known Rosenbrock (1960) banana-shaped valley problem with the methods of this chapter. We will start from the "standard" point of $(-1.2, 1)$, and we will not, for the moment, supply the gradient.

```
fr <- function(x) {
    ## Rosenbrock Banana function
    x1 <- x[1]
    x2 <- x[2]
    100 * (x2 - x1 * x1)^2 + (1 - x1)^2
}
sstart <- c(-1.2, 1)
```

To keep the output tidy, we will use a little display function.

```
prtopt <- function(optres, headname) {
    # compact print of optim() result
    cat(headname, ": f(")
    npar <- length(optres$par)
    for (i in 1:(npar - 1)) {
        cat(optres$par[[i]], ", ")
    }
    cat(optres$par[[npar]], ") = ", optres$value, " after ",
        optres$counts[[1]], "f & ", optres$counts[[2]], "g\n")
    cat(optres$message, " convergence code=", optres$convergence,
        "\n")
}
```

First, we try the Nelder–Mead, BFGS, CG, and L-BFGS-B methods of `optim()`

```
anm <- optim(sstart, fr) # Nelder Mead is the default method
prtopt(anm, "anm")

## anm : f(1.0003 , 1.0005 ) =  8.8252e-08  after   195 f &  NA g
##     convergence code= 0

abfgs <- optim(sstart, fr, method = "BFGS")
prtopt(abfgs, "abfgs")

## abfgs : f(0.9998 , 0.99961 ) =  3.8274e-08  after  118 f &  38g
##     convergence code= 0

acg1 <- optim(sstart, fr, method = "CG", control = list(type = 1))
prtopt(acg1, "acg1")

## acg1 : f(-0.76481 , 0.59271 ) =  3.1065  after  402 f &  101g
##     convergence code= 1

acg2 <- optim(sstart, fr, method = "CG", control = list(type = 2))
prtopt(acg2, "acg2")

## acg2 : f(0.96767 , 0.93628 ) =  0.0010432  after  384 f &  101g
##     convergence code= 1
```

```
acg3 <- optim(sstart, fr, method = "CG", control = list(type = 3))
prtopt(acg3, "acg3")

## acg3 : f(0.94331 , 0.88953 ) =   0.0032119  after   352 f &   101g
##     convergence code= 1

albfgsb <- optim(sstart, fr, method = "L-BFGS-B")
prtopt(albfgsb, "albfgsb")

## albfgsb : f(0.9998 , 0.9996 ) =   3.9985e-08  after   49 f &   49 g
## CONVERGENCE: REL_REDUCTION_OF_F <= FACTR*EPSMCH    convergence code= 0
```

Here we note that Nelder–Mead, BFGS, and L-BFGS-B have essentially found the correct solution at $(1, 1)$, but the CG examples have hit the maximum allowed gradient computations. Let us check the results using the analytic gradient. Note that we test the gradient code!

```
frg <- function(x) {
    ## Rosenbrock Banana function gradient
    x1 <- x[1]
    x2 <- x[2]
    g1 <- -400 * (x2 - x1 * x1) * x1 - 2 * (1 - x1)
    g2 <- 200 * (x2 - x1 * x1)
    gg <- c(g1, g2)
}

## test gradient at standard start:
require(numDeriv)
gn <- grad(fr, sstart)
ga <- frg(sstart)
## max(abs(ga-gn))
max(abs(ga - gn))

## [1] 1.47e-08

## analytic gradient results
gabfgs <- optim(sstart, fr, frg, method = "BFGS")
prtopt(gabfgs, "BFGS w. frg")

## BFGS w. frg : f(1 , 1 ) =   9.595e-18  after   110 f &   43g
##     convergence code= 0

gacg1 <- optim(sstart, fr, frg, method = "CG", control = list(type = 1))
prtopt(gacg1, "CG Fletcher-Reeves w. frg")

## CG Fletcher-Reeves w. frg : f(-0.76484,0.59276) = 3.1066 after 402 f & 101g
##     convergence code= 1

gacg2 <- optim(sstart, fr, frg, method = "CG", control = list(type = 2))
prtopt(gacg2, "CG Polak-Ribiere w. frg")

## CG Polak-Ribiere w. frg : f(0.99441,0.98882)=3.1238e-05 after 385 f & 101g
##     convergence code= 1
```

```
gacg3 <- optim(sstart, fr, frg, method = "CG", control = list(type = 3))
prtopt(gacg3, "CG Beale-Sorenson w. frg")

## CG Beale-Sorenson w. frg : f(0.89973,0.80901) = 0.010045 after 355 f & 101g
##    convergence code= 1

galbfgsb <- optim(sstart, fr, method = "L-BFGS-B")
prtopt(galbfgsb, "L-BFGS-B w. frg")

## L-BFGS-B w. frg : f(0.9998 , 0.9996 ) = 3.9985e-08 after  49 f &  49 g
## CONVERGENCE: REL_REDUCTION_OF_F <= FACTR*EPSMCH   convergence code= 0
```

We observe that the CG methods are still doing poorly compared to other approaches.

References

Dai YH and Yuan Y 2001 An efficient hybrid conjugate gradient method for unconstrained optimization. Annals of Operations Research 103(1–4), 33–47.

Fletcher R 1970 A new approach to variable metric algorithms. Computer Journal 13(3), 317–322.

Fox PA, Hall AD and Schryer NL 1978 The PORT mathematical subroutine library. ACM Transactions on Mathematical Software (TOMS) 4(2), 104–126.

Lu P, Nocedal J, Zhu C, Byrd RH and Byrd RH 1994 A limited-memory algorithm for bound constrained optimization. SIAM Journal on Scientific Computing 16, 1190–1208.

Morales JL and Nocedal J 2011 Remark on Algorithm 778: L-BFGS-B: fortran subroutines for large-scale bound constrained optimization. ACM Transactions on Mathematical Software 38(1), 7:1–7:4.

Nash JC 1979 Compact Numerical Methods for Computers: Linear Algebra and Function Minimisation. Hilger, Bristol.

Nash JC 2000 The Dantzig simplex method for linear programming. Computing in Science and Engineering 2(1), 29–31.

Nelder JA and Mead R 1965 A simplex method for function minimization. Computer Journal 7(4), 308–313.

Rosenbrock HH 1960 An automatic method for finding the greatest or least value of a function. Computer Journal 3, 175–184.

Schnabel RB, Koonatz JE and Weiss BE 1985 A modular system of algorithms for unconstrained minimization. ACM Transactions on Mathematical Software 11(4), 419–440.

Zhu C, Byrd RH, Lu P and Nocedal J 1997 Algorithm 778: L-bfgs-b: fortran subroutines for large-scale bound-constrained optimization. ACM Transactions on Mathematical Software 23(4), 550–560.

9

Add-in function minimization packages for R

CRAN (Comprehensive R Archive Network) and R-Forge have a number of packages for function minimization. Indeed, I am a developer involved in several of them. This chapter explores some of these. Truthfully, there is more material than can be sensibly included in one book, let alone one chapter of a book, but we will do our best to provide a decent overview.

9.1 Package `optimx`

The package `optimx` (Nash and Varadhan, 2011a, 2011b) is a wrapper package for a number of optimization routines, including several of those included in the base system. One purpose is to unify the calls to a number of useful optimizers into a common syntax and thereby allow their performance comparison in an easy way. This ability for comparison makes the package a useful introduction to a number of the packages it calls. Different syntaxes for the calls to different routines are a great nuisance for the user and a frequent source of error.

However, `optimx` is more than simply a tool to unify the calling sequence syntax. The design incorporates several important concepts that are discussed elsewhere in this book (see Chapter 3).

- A number of initial tests of the function, gradient, and scaling are available before optimization is attempted.

- After optimizers are complete, Kuhn–Karush–Tucker verification of the solutions is the default behavior if the number of parameters is not too large.

- `summary()`, `coef()`, and `print()` methods are available for the returned results.

Nonlinear Parameter Optimization Using R Tools, First Edition. John C. Nash.
© 2014 John Wiley & Sons, Ltd. Published 2014 by John Wiley & Sons, Ltd.
Companion Website: www.wiley.com/go/nonlinear_parameter

There are other features that could improve the success of optimization calculations, and these have been investigated in the experimental package **opt-plus** that is in the optimizer project on R-forge (https://r-forge.r-project.org/R/?group_id=395). This extends **optimx** as follows:

- Parameter scaling, as in the optim() function, is available for all methods. This extension added a great deal of complexity to the code and incurs a performance penalty.

- Similarly, function evaluations are checked to see if they are computable, so that trial points or search directions can be revised. This allows optimization to proceed and avoids some stoppages that kill a user's program, but for problems where trial points are all computable is a waste of time.

- After optimizers complete, an axial search verification of the solutions is the default behavior. This requires $2 * n$ function evaluations for each method, where n is the number of parameters.

Unfortunately, the performance costs of these features appear to outweigh the reliability benefits. Scaling and function checking are useful ideas, but I have yet to discover a general but efficient implementation in R. When an optimization task is being carried out many times, we would like to separate out some of the checks and scaling so that they are not repeated unnecessarily. For performance, the scaling and checks should be applied at the stage that the objective and gradient functions are written. On the other hand, when we want to solve a problem only one time, such checks can help to avoid time-consuming errors and scaling can help us to find a solution.

9.1.1   Optimizers in optimx

At the time of writing, the optimization methods in **optimx** includes the following methods from the base system that were discussed in Chapter 8:

- the BFGS, Nelder–Mead, L-BFGS-B, and CG methods from the optim() function;

- nlm();

- nlminb().

Also included (again at time of writing) are

- package **Rvmmin** (Nash, 2011b), which is my all-R implementation of the variable metric method of Fletcher (1970);

- package **Rcgmin** (Nash, 2011a), which is my all-R implementation of the conjugate gradient method of Dai and Yuan (2001);

- function spg from **BB** (Varadhan and Gilbert, 2009), an R implementation of the Birgin and Raydan (2001) code by Ravi Varadhan;

- package **ucminf** (Nielsen and Mortensen, 2012), a variable metric method using an inverse Hessian approximation with a line search due to Nielsen (2000);

- functions nmkb and hjkb from package **dfoptim** (Varadhan and Borchers, 2011), which are Kelley's (1987) variant of the Nelder–Mead method with bounds and a Hooke and Jeeves (1961) method with bounds

- methods newuoa and bobyqa from package **minqa** (Bates et al. 2010), an interfacing of Powell's BOBYQA Powell (2009) that is written in Fortran;

9.1.2 Example use of optimx()

Let us use the traditional Rosenbrock function fr() with, where it can be used, the analytic gradient frg() as presented in Chapter 8.

```
require(optimx)

## Loading required package: optimx

fr <- function(x) {
    ## Rosenbrock Banana function
    x1 <- x[1]
    x2 <- x[2]
    100 * (x2 - x1 * x1)^2 + (1 - x1)^2
}
frg <- function(x) {
    ## Rosenbrock Banana function gradient
    x1 <- x[1]
    x2 <- x[2]
    g1 <- -400 * (x2 - x1 * x1) * x1 - 2 * (1 - x1)
    g2 <- 200 * (x2 - x1 * x1)
    gg <- c(g1, g2)
}
sstart <- c(-1.2, 1)
arb <- optimx(par = sstart, fn = fr, gr = frg, method = "all")

## Loading required package: numDeriv
## Warning: bounds can only be used with method L-BFGS-B (or Brent)
## Warning: bounds can only be used with method L-BFGS-B (or Brent)
## Warning: bounds can only be used with method L-BFGS-B (or Brent)

print(summary(arb, order = value))

##                 p1     p2     value fevals gevals niter
## Rvmmin      1.0000 1.0000 1.700e-26     59     39    NA
## nlminb      1.0000 1.0000 4.292e-22     43     36    35
## Rcgmin      1.0000 1.0000 8.805e-18    735    434    NA
## ucminf      1.0000 1.0000 2.726e-17     38     38    NA
## newuoa      1.0000 1.0000 2.178e-15    257     NA    NA
## BFGS        1.0000 1.0000 2.268e-13     47     47    NA
## CG          1.0000 1.0000 2.268e-13     47     47    NA
## Nelder-Mead 1.0000 1.0000 2.268e-13     47     47    NA
```

```
## L-BFGS-B    1.0000 1.0000 2.268e-13      47     47    NA
## bobyqa      1.0000 1.0000 2.845e-13     290     NA    NA
## nlm         1.0000 1.0000 3.974e-12      NA     NA    23
## hjkb        1.0000 0.9999 2.618e-09     540     NA    19
## spg         0.9998 0.9996 3.896e-08     141     NA   112
## nmkb        1.0003 1.0006 4.269e-07     120     NA    NA
##             convcode  kkt1 kkt2 xtimes
## Rvmmin            2   TRUE TRUE  0.004
## nlminb            0   TRUE TRUE  0.000
## Rcgmin            0   TRUE TRUE  0.032
## ucminf            0   TRUE TRUE  0.000
## newuoa            0   TRUE TRUE  0.000
## BFGS              0   TRUE TRUE  0.000
## CG                0   TRUE TRUE  0.000
## Nelder-Mead       0   TRUE TRUE  0.000
## L-BFGS-B          0   TRUE TRUE  0.000
## bobyqa            0   TRUE TRUE  0.004
## nlm               0   TRUE TRUE  0.000
## hjkb              0   TRUE TRUE  0.008
## spg               0   TRUE TRUE  0.060
## nmkb              0  FALSE TRUE  0.004
```

optimx() allows a very quick comparison of how well methods work on particular classes of problem. It should not, of course, be used in this way for obtaining solutions on a production basis. In the present example, we see that we could use any of the first half dozen methods satisfactorily.

9.2   Some other function minimization packages

The collection of packages for R is large and fluid. Those in what I consider the primary collection, the Comprehensive R Archive Network or CRAN, must satisfy regular build tests, so sometimes disappear from the collection. Overall, the size of the collection is growing. There are, moreover, many other repositories and collections, some are excellent and others are of dubious value. Thus my comments here are directed to some tools that I consider useful.

9.2.1   nloptr and nloptwrap

nloptr (Ypma, 2013) is an interface to the NLopt library maintained by Steven G. Johnson of MIT (http://ab-initio.mit.edu/wiki/index.php/NLopt). There is a simplified wrapper for this called nloptwrap (Borchers, 2013). There are a number of optimization tools in NLopt, some of which mirror, and are derived from the same sources as, some of the tools seen above for unconstrained and bounds constrained optimization.

```
require(nloptwrap)

## Loading required package: nloptwrap
## Loading required package: nloptr
##
## Attaching package: 'nloptwrap'
##
```

```
## The following objects are masked from 'package:minqa':
##
##    bobyqa, newuoa

nowp <- function(answer) {
    # nloptwrap answer summary
    cat("Fn =", answer$value, " after ", answer$iter, " iterations,
 parameters:\n")
    print(answer$par)
    cat(answer$message, "\n")
    invisible(0)
}

albfgs <- lbfgs(sstart, fr, gr = frg)
nowp(albfgs)

## Fn = 7.357e-23  after  56  iterations, parameters:
## [1] 1 1
## NLOPT_SUCCESS: Generic success return value.

atnewton <- tnewton(sstart, fr, gr = frg)
nowp(atnewton)

## Fn = 1.735e-24  after  74  iterations, parameters:
## [1] 1 1
## NLOPT_SUCCESS: Generic success return value.

avarmetric <- varmetric(sstart, fr, gr = frg)
nowp(avarmetric)

## Fn = 1.391e-18  after  53  iterations, parameters:
## [1] 1 1
## NLOPT_SUCCESS: Generic success return value.

anelmead <- neldermead(sstart, fr)
nowp(anelmead)

## Fn = 3.656e-14  after  214  iterations, parameters:
## [1] 1 1
## NLOPT_XTOL_REACHED: Optimization stopped because xtol_rel or
 xtol_abs (above) was reached.

anewuoa <- newuoa(sstart, fr)
nowp(anewuoa)

## Fn = 8.023e-17  after  194  iterations, parameters:
## [1] 1 1
## NLOPT_SUCCESS: Generic success return value.

# remove the packages in case of confusion
detach(package:nloptwrap)
detach(package:nloptr)
```

9.2.2 `trust` and `trustOptim`

Trust region methods of optimization, as pointed out very nicely in http://en.
wikipedia.org/wiki/Trust_region, choose a step size for a search for a new and
hopefully better set of parameters and then find a search direction. The step size is

determined by the region in which a model of the objective function is sufficiently accurate for the needs of optimization. Traditional gradient methods choose the search direction and then work out a step size.

trust (Geyer, 2013) implements a trust region method where the user must supply objective function code and a starting set of parameters. However, the objective function now returns not only the function but also its gradient and Hessian. This is, for many problems, a large and onerous task. In the following text, we illustrate this using the Rosenbrock function that has just been used above. Furthermore, we use the example code from **trust** to show an alternative way to generate the objective function in objfun2().

```
frh <- function(x) {
    ## Rosenbrock Banana function gradient
    x1 <- x[1]
    x2 <- x[2]
    h11 <- -400 * x2 + 1200 * x1 * x1 + 2
    h12 <- -400 * x1
    h21 <- h12
    h22 <- 200
    HH <- matrix(c(h11, h12, h21, h22), nrow = 2, ncol = 2)
}
objfun1 <- function(x) {
    val <- fr(x)
    gg <- frg(x)
    HH <- frh(x)
    list(value = val, gradient = gg, hessian = HH)
}
objfun2 <- function(x) {
    stopifnot(is.numeric(x))
    stopifnot(length(x) == 2)
    f <- expression(100 * (x2 - x1^2)^2 + (1 - x1)^2)
    g1 <- D(f, "x1")
    g2 <- D(f, "x2")
    h11 <- D(g1, "x1")
    h12 <- D(g1, "x2")
    h22 <- D(g2, "x2")
    x1 <- x[1]
    x2 <- x[2]
    f <- eval(f)
    g <- c(eval(g1), eval(g2))
    B <- rbind(c(eval(h11), eval(h12)), c(eval(h12), eval(h22)))
    list(value = f, gradient = g, hessian = B)
}
require(trust)
atrust1 <- trust(objfun1, sstart, rinit = 1, rmax = 5)
atrust1

## $value
## [1] 3.7902e-22
##
## $gradient
## [1]  2.9933e-10 -1.6724e-10
##
## $hessian
##        [,1] [,2]
## [1,]  802 -400
```

```
## [2,] -400  200
##
## $argument
## [1] 1 1
##
## $converged
## [1] TRUE
##
## $iterations
## [1] 27

## Get same answers from atrust2<-trust(objfun2, sstart,
## rinit=1, rmax=5)
```

We note that **trust** works fine for the Rosenbrock function, which it should, given the amount of information provided.

Where should a package such as **trust** be used? As it requires the gradient and Hessian as well as the function to be computed, it clearly will not be advisable to apply it to functions where these are difficult or impossible to compute. This includes functions where we have conditional statements, absolute values, or similar constructs. Moreover, **trust** will be most favored when the gradient and Hessian are easy to compute, which can be the case for certain likelihoods based on exponential family probability distributions. Communications from its author, Charles Geyer, make it clear that he understands this very well and that **trust** is intended for situations where it is very economical to compute the objective function, gradient, and Hessian in a coordinated manner.

If gradient and hessian functions are available, we still need to consider the control parameters `rinit` and `rmax`. These values are likely not critical, but defaults are not provided.

A more recent package is **trustOptim** (Braun, 2013), which claims to be usable for sparse Hessians. There is a vignette to illustrate how this package can be used. However, some suggested test cases supplied by the package author gave unexpected results, so there are uncertainties about the package at the time of writing. I suspect that in time this package will become useful in the sphere of application for which it is intended, but at the time of writing, urge caution in using it.

9.3   Should we replace `optim()` routines?

In several places in this book, I have noted that some of the methods in `optim()` are "old," even though they are my own. Among those of us working to improve the optimization tools in R, there is an ongoing discussion of how to replace the minimizers in `optim()`. This is, and should be, an ongoing task and discussion. With **optimx**, we do have some reasonable and indeed evolving replacements for the methods in `optim()`. However, there remains an important and possibly difficult adjustment to be made in various packages and functions that make use of `optim()`, in particular those that directly call the internal functions that make up the `optim()` function. For example, `nnet()` directly calls the internal C routine vmmin.c which is the principal computational engine of the optim() method

"BFGS." The difficulty is less in offering "new" functions than in removing the obsolete ones.

References

Bates D, Mullen KM, Nash JC and Varadhan R 2010 **minqa**: Derivative-Free Optimization Algorithms by Quadratic Approximation. R Foundation for Statistical Computing. R package version 1.1.13.

Birgin EG and Raydan MM 2001 Spg: software for convex-constrained optimization. ACM Transactions on Mathematical Software 27, 340–349.

Borchers HW 2013 *nloptwrap: Wrapper for Package nloptr*. R package version 0.5-1.

Braun M 2013 *trustOptim: Trust region nonlinear optimization, efficient for sparse Hessians*. R package version 0.8.1.

Dai YH and Yuan Y 2001 An efficient hybrid conjugate gradient method for unconstrained optimization. Annals of Operations Research 103(1–4), 33–47.

Fletcher R 1970 A new approach to variable metric algorithms. Computer Journal 13(3), 317–322.

Geyer CJ 2013 *trust: Trust Region Optimization*. R package version 0.1-4.

Hooke R and Jeeves TA 1961 "direct search" solution of numerical and statistical problems. Journal of the ACM 8(2), 212–229.

Kelley C 1987 *Iterative Methods for Optimization* Frontiers in Applied Mathematics. Society for Industrial and Applied Mathematics.

Nash JC 2011a **Rcgmin**: *conjugate Gradient Minimization of Nonlinear Functions with Box Constraints*. Nash Information Services Inc. R package version 2011-2.10.

Nash JC 2011b **Rvmmin**: *variable Metric Nonlinear Function Minimization with Bounds Constraints*. Nash Information Services Inc. R package version 2011-2.25.

Nash JC and Varadhan R 2011a **optimx**: *a Replacement and Extension of the optim() function*. Nash Information Services Inc. and Johns Hopkins University. R package version 2011-3.5.

Nash JC and Varadhan R 2011b Unifying optimization algorithms to aid software system users: optimx for R. Journal of Statistical Software 43(9), 1–14.

Nielsen HB 2000 UCMINF - an algorithm for unconstrained, nonlinear optimization. Technical report, Department of Mathematical Modelling, Technical University of Denmark. Report IMM-REP-2000-18.

Nielsen HB and Mortensen SB 2012 *ucminf: General-purpose unconstrained non-linear optimization*. R package version 1.1-3.

Powell MJD 2009 The BOBYQA algorithm for bound constrained optimization without derivatives.

Varadhan R and Borchers HW 2011 *dfoptim: Derivative-free Optimization* Johns Hopkins University and ABB Corporate Research. R package version 2011.12-9.

Varadhan R and Gilbert P 2009 **BB**: an R package for solving a large system of nonlinear equations and for optimizing a high-dimensional nonlinear objective function. Journal of Statistical Software 32(4), 1–26.

Ypma J 2013 *nloptr: R interface to NLopt*. R package version 0.9.3.

10

Calculating and using derivatives

In previous chapters, we have referred several times to derivatives, that is, to gradients and Hessians of functions. In practice, having good derivative information is important to obtaining solutions or to knowing that we have a valid solution. This chapter will look at ways in which we can acquire and use such information.

10.1 Why and how

Derivative information is important

- because the KKT conditions (Karush, 1939; Kuhn and Tucker, 1951) for a minimum require first derivatives to be "zero" and the Hessian, that is, second derivative matrix, to be positive definite;

- because many methods can use gradient information.

Indeed, even methods that claim to be "derivative free" will often use the concepts of gradients and Hessians, either for the function to be minimized or for an approximating model.

It is my experience that the main utility of good derivative information is in testing that we indeed have a solution. That is, it is useful for termination test and improves performance because it allows us to cease trying to proceed when our journey is complete. In some cases, approximate derivatives may actually give better performance for some gradient methods in initial steps when we are far from the solution. This is similar to secant methods outperforming Newton methods in early iterations.

Nonlinear Parameter Optimization Using R Tools, First Edition. John C. Nash.
© 2014 John Wiley & Sons, Ltd. Published 2014 by John Wiley & Sons, Ltd.
Companion Website: www.wiley.com/go/nonlinear_parameter

Unfortunately, the calculation of derivatives is not a trivial task. This chapter looks at some approaches and presents some recommendations that I freely admit are open to discussion –they are my own views based on over four decades of dealing with these matters, but they are inevitably grounded in the particular problems I have had to try to solve, which may be very different from the problems faced by my readers.

The following are some general considerations and suggestions.

- Always run a check on the computations. Indeed, some of the packages I have helped develop include checks of derivatives against numerical approximations because derivative coding is so error prone.

- Comment your code. I usually find the same person who wrote the code (me!) is the one who has to figure it out later on, so it is important for my own working efficiency to try to write readable scripts. In this respect, derivatives can be particularly tricky.

- If there are bounds, or if the function is undefined in some areas of the domain, then try to include proper checks to avoid a "crash" or, much worse, getting results that cause a silent failure, that is, where the answer is wrong but there is no error message.

10.2   Analytic derivatives – by hand

For problems that are run frequently, I believe that manually coded derivative functions generally allow the best performance. If the programmer is sensible, they will also be commented and more comprehensible to the reader of the code than any automatically generated function. Note that for some functions, it can be quite efficient to compute the objective function, the gradient and the Hessian together, because there may be common subcomputations.

Unfortunately, working out analytic derivatives is tedious and error prone. For nonlinear least squares or similar computations where one is trying to fit a function model(parms) to data by varying the parameters parms, I have found it useful to write residuals in the form

$$resids(parms) = model(parms) - data \qquad (10.1)$$

This avoids one possible sign error in taking the derivatives of the resids, even though most statisticians will write the residuals in the data – model(parms) form. Of course, it makes no difference to the sum of squares. I urge readers to develop similar tricks that suit their own working habits so they can to avoid simple but costly errors in computing derivatives. As noted by McCullagh and Nelder (1989, p. 9), both Gauss and Legendre use the form above for residuals, so I am in good company.

10.3   Analytic derivatives – tools

As mentioned, there are software tools that permit symbolic mathematics and these could be used to generate expressions for the derivatives. I have myself made occasional use of such packages, for example, DERIVE (Nash, 1995). (This appears to be no longer generally available.) I have also made use on occasions of Maxima (http://maxima.sourceforge.net). However, I find that these tools are quite difficult to use well and need ongoing use to maintain one's skills with them. If you are willing to use these packages regularly, I believe that they may offer considerable rewards.

A particular use of symbolic maths tools is in allowing for partial development of some of the expressions used in the "by-hand" coding of derivative functions. They may also provide a way of checking that our code generates correct answers, possibly for intermediate stages of the derivative computation.

Similarly, there are packages for automatic differentiation (AD). This is not the same as symbolic differentiation, where a computer program is provided with a symbolic representation of the function to be differentiated and returns a symbolic expression with the derivative(s). Instead, clever application of the chain rule is used to compute a numerical value of the derivative(s) wanted, with the "function" supplied as computational code. Thus AD tools must be adapted to the language in which functions are written. In our case, this is R.

R offers some tools that combine symbolic differentiation and AD, and for readers of this book, I believe that these provide the most value for effort expended. However, the tools are limited in their scope. That is, they encompass only a limited set of functions for which derivatives can be computed.

10.4   Examples of use of R tools for differentiation

Let us use our nonlinear least squares Example 1.2. We set up the model expression in mod, provide the parameter names in namev, and use the R function deriv() to get the expression for the gradient.

```
mod <- ~100 * b1/(1 + 10 * b2 * exp(-0.1 * b3 * t))
mod

## ~100 * b1/(1 + 10 * b2 * exp(-0.1 * b3 * t))

namev <- c("b1", "b2", "b3")
try1 <- deriv(mod, namev)
try1

## expression({
##     .expr1 <- 100 * b1
##     .expr2 <- 10 * b2
##     .expr6 <- exp(-0.1 * b3 * t)
```

```
##    .expr8 <- 1 + .expr2 * .expr6
##    .expr13 <- .expr8^2
##    .value <- .expr1/.expr8
##    .grad <- array(0, c(length(.value), 3L), list(NULL, c("b1",
##        "b2", "b3")))
##    .grad[, "b1"] <- 100/.expr8
##    .grad[, "b2"] <- -(.expr1 * (10 * .expr6)/.expr13)
##    .grad[, "b3"] <- .expr1 * (.expr2 * (.expr6 * (0.1 * t)))/.expr13
##    attr(.value, "gradient") <- .grad
##    .value
## })
```

We can now copy and paste this expression into a function for computing the residuals and Jacobian of the model. We then compute the sum of squares and gradient.

```
resfn <- function(pars, t = t, y = y) {
    b1 <- pars[[1]]
    b2 <- pars[[2]]
    b3 <- pars[[3]]
    .expr1 <- 100 * b1
    .expr2 <- 10 * b2
    .expr6 <- exp(-0.1 * b3 * t)
    .expr8 <- 1 + .expr2 * .expr6
    .expr13 <- .expr8^2
    .value <- .expr1/.expr8
    .grad <- array(0, c(length(.value), 3L), list(NULL, c("b1",
        "b2", "b3")))
    .grad[, "b1"] <- 100/.expr8
    .grad[, "b2"] <- -(.expr1 * (10 * .expr6)/.expr13)
    .grad[, "b3"] <- .expr1 * (.expr2 * (.expr6 * (0.1 * t)))/.expr13
    attr(.value, "gradient") <- .grad
    .value
    res <- .value - y
    attr(res, "jacobian") <- .grad
    res
}
## Test the function
ydat <- c(5.308, 7.24, 9.638, 12.866, 17.069, 23.192, 31.443,
    38.558, 50.156, 62.948, 75.995, 91.972) # for testing
tdat <- 1:length(ydat) # for testing
start <- c(1, 1, 1)
myres <- resfn(start, y = ydat, t = tdat)
print(as.vector(myres)) # we don't want the 'gradient'

##  [1]   4.64386   3.64458   2.25518   0.11562  -2.91533
##  [6]  -7.77921 -14.68094 -20.35397 -30.41538 -41.57497
## [11] -52.89343 -67.04642

ss <- as.numeric(crossprod(as.vector(myres)))
ss

## [1] 10685

JJ <- attr(myres, "jacobian")
JJ
```

```
##               b1         b2         b3
## [1,]    9.9519   -8.9615   0.89615
## [2,]   10.8846   -9.6998   1.93997
## [3,]   11.8932  -10.4787   3.14361
## [4,]   12.9816  -11.2964   4.51856
## [5,]   14.1537  -12.1504   6.07520
## [6,]   15.4128  -13.0373   7.82235
## [7,]   16.7621  -13.9524   9.76668
## [8,]   18.2040  -14.8902  11.91213
## [9,]   19.7406  -15.8437  14.25933
## [10,]  21.3730  -16.8050  16.80496
## [11,]  23.1016  -17.7647  19.54122
## [12,]  24.9256  -18.7127  22.45528
```

```
svd(JJ)$d
```

```
## [1] 86.61282 12.82005  0.33299
```

```
grad <- crossprod(JJ, as.vector(myres))
print(as.vector(grad))
```

```
## [1] -5045.7  3917.7 -4117.1
```

Note that we need to get the individual parameters out of the vector pars or else change the expressions inside the gradient evaluation to extract these values directly. While I believe that it should not be too difficult to use R to automate the "copy and paste" and make the appropriate changes, I am not convinced that it is worth the effort.

10.5   Simple numerical derivatives

We can approximate derivatives by considering the elementary calculus expressions from which limits yield differentiation. Let us for the moment consider $f(x)$, a function of a single parameter x. That is, using a forward step of h, an approximation to the derivative is

$$d_{\text{forward}} = \{f(x+h) - f(x)\} / h \qquad (10.2)$$

The only awkward issue is the choice of the stepsize h. In calculus, the limit sends this to zero. In practice, we want it to be big enough that $f(x + h)$ has enough digits different from $f(x)$ to allow a good approximation to the true derivative. However, if it is too large, then we may encounter larger scale changes in $f(x)$. As a compromise, I have found that a value computed as

$$h = \text{sqrt(eps)} * (\text{abs}(x) + \text{sqrt(eps)}) \qquad (10.3)$$

where eps is the floating point precision – the smallest number such that $1 + \text{eps}$ tests greater than 1 – is a reasonable compromise (Nash, 1979, Chapter 18), but no value can be regarded as universally satisfactory.

There is, of course, no particular reason for stepping forward, and a backward derivative approximation is equally reasonable:

$$d_{\text{backward}} = \{f(x) - f(x - h)\}/h \qquad (10.4)$$

A central approximation averages these:

$$d_{\text{central}} = 0.5 * (\{f(x) - f(x - h)\}/h + \{f(x + h) - f(x)\}/h)$$
$$= \{f(x + h) + f(x - h) - 2 * f(x)\}/(2 * h)$$

As Wikipedia points out (http://en.wikipedia.org/wiki/Finite_difference), the central difference yields an approximation with error proportional to the square of h, while for forward and backward approximations the error is proportional to h. We pay for the extra accuracy with an extra function evaluation. The central difference approximation is the default in function `grad()` from package **pracma** (Borchers, 2013). Note that there is also a `grad()` in package **numDeriv**.

10.6   Improved numerical derivative approximations

There are a couple of other ways by which we can improve our numerical approximations.

10.6.1   The Richardson extrapolation

The Richardson extrapolation is a general technique that aims to estimate a "better" answer for approximate computations. The forward, backward, and central derivative approximations of the previous section all become the desired derivative in the limit of h going to zero, but only if we can evaluate the function sufficiently precisely, so the subtraction does not cancel all the digits. The idea of the Richardson extrapolation is to compute the derivative approximation for several values of h and extrapolate to get the value for $h = 0$. The same idea when extended to numerical integration is called the Romberg method. Wikipedia provides a readable summary (http://en.wikipedia.org/wiki/Richardson_extrapolation.

R package **numDeriv** (Gilbert, 2009) uses the Richardson extrapolation as its default method for computing derivative approximations. `method="simple"` or `method="complex"` can also be specified.

10.6.2   Complex-step derivative approximations

If the function for which we wish derivatives is analytic in complex values of the function, it turns out that we can find approximations that are almost as good as the analytic derivatives. This somewhat surprising result seems to have been first published by James Lyness. Besides many contributions to numerical mathematics, I owe James a personal debt of gratitude for saving me from an uncomfortable night at O'Hare Airport when a careless equipment operator cut a power cable for runway lights. A short and readable account of complex-step derivatives is given by Squire and Trapp (1998).

The complex-step approximation to the first derivative of $f(x)$ is given as

$$f'(x) \sim \mathrm{Im}(f(x + ih))/h \qquad (10.5)$$

This requires only one function evaluation and can be shown to provide a very accurate approximation when it works. The fly in the ointment is that we need a function that can generate a complex result. Thus anything with "if"-like structures, absolute values, or similar forms will fail. Generally, it is best to stick to very simple functions when using this derivative approximation, although it is certainly worth trying. Here is a small example using **numDeriv**. Thanks to Hans Werner Borchers for this.

```
detach(package:pracma) # avoid confusion between two grad() functions
f1 <- function(x) 1 + x^2
f2 <- function(x) abs(1 + x^2)
require(numDeriv)
grad(f1, 1, method = "complex")

## [1] 2

grad(f2, 1, method = "complex")

## Error: function does not return a complex value as required by method
  'complex'.
```

10.7 Strategy and tactics for derivatives

Before suggesting what I believe are reasonable approaches for derivative calculations, let us examine the costs and accuracy of different methods (Figure 10.1). We will use the generalized Rosenbrock function with scale factor 100 for different numbers of parameters.

```
genrose.f <- function(x, gs = 100) {
    # objective function A generalization of the Rosenbrock
    # banana valley function to n parameters
    n <- length(x)
    fval <- 1 + sum(gs * (x[1:(n - 1)]^2 - x[2:n])^2 + (x[2:n] -
        1)^2)
    return(fval)
}

genrose.g <- function(x, gs = 100) {
    # vectorized gradient for genrose.f
    n <- length(x)
    gg <- as.vector(rep(0, n))
    tn <- 2:n
    tn1 <- tn - 1
    z1 <- x[tn] - x[tn1]^2
```

```
z2 <- 1 - x[tn]
gg[tn] <- 2 * (gs * z1 - z2)
gg[tn1] <- gg[tn1] - 4 * gs * x[tn1] * z1
return(gg)
}
```

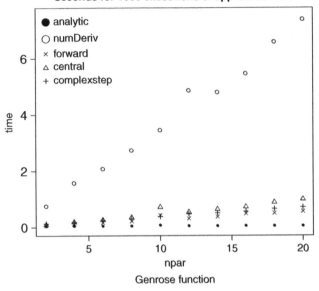

Seconds for 1000 executions of approximations

Figure 10.1

We compute relative errors in the gradients compared to analytic values by taking the ratio of the maximum absolute error to the maximum absolute gradient
element. That is, we are not computing all the relative errors and then finding their
maximum absolute value. This is to avoid a zero divide, while still obtaining a
measure of the relative error. The results are given in the table at the end of this
section.

Overall, my recommendation to users of optimization software is to try first to
use analytic derivatives. This may sometimes give slightly slower timings or work
measures for the optimization calculation than using a numerical approximation
to the derivative, especially when the starting point is far from the solution. At
least a partial understanding of this reality can be seen from an one-dimensional
problem, where a secant line will in early iterations make better progress toward
a minimum than the Newton (i.e., tangent) line. On the other hand, the analytic
derivatives are much more reliable for determining that we are actually at an optimum. Moreover, they save much work in running the KKT and other optimality
tests, as well as in evaluating test quantities that are used to decide termination of
an optimization code.

If analytic derivatives are not available, then clearly the complex- step approximation, which here gives full accuracy, is the next best choice, but it may not be

computable for particular functions. While **numDeriv** does better than a central difference approximation, the latter is much more economical of computing time and does achieve a reasonable accuracy.

Derivative relative errors

```
##                                  ##
## ------------------------------------------------------------
## npar   forward   central   numDeriv   complexstep
## ------  ---------  ---------  ---------  -----------
##  2    3.25e-07   7.57e-10   6.51e-12        0
##                                  ##
##  4    3.53e-07   8.37e-10   6.21e-12        0
##                                  ##
##  6    3.53e-07   8.37e-10   9.91e-12        0
##                                  ##
##  8    3.53e-07   5.37e-10   2.49e-12        0
##                                  ##
## 10    3.52e-07   1.18e-09   1.11e-11        0
##                                  ##
## 12    3.53e-07   8.37e-10   2.49e-12        0
##                                  ##
## 14    3.55e-07   8.37e-10   1.07e-11        0
##                                  ##
## 16    3.52e-07   1.18e-09   6.94e-12        0
##                                  ##
## 18    3.52e-07   1.18e-09   7.84e-12        0
##                                  ##
## 20    3.52e-07   1.18e-09   1.86e-11        0
## ------------------------------------------------------------
```

References

Borchers HW 2013 pracma: Practical Numerical Math Functions. R package version 1.5.0.

Gilbert P 2009 **numDeriv**: Accurate Numerical Derivatives. R Foundation for Statistical Computing. R package version 2009.2-1.

Karush W 1939 Minima of functions of several variables with inequalities as side constraints. Master's thesis, MSc Dissertation, Department of Mathematics, University of Chicago Chicago, IL.

Kuhn HW and Tucker AW 1951 Nonlinear programming. Proceedings of 2nd Berkeley Symposium, pp. 481–492, Berkeley, USA.

McCullagh P and Nelder J 1989 Generalized Linear Models, Monographs on Statistics and Applied Probability, Second Edition. Chapman and Hall, London.

Nash JC 1979 Compact Numerical Methods for Computers: Linear Algebra and Function Minimisation. Adam Hilger, Bristol. Second Edition, 1990, Bristol: Institute of Physics Publications.

Nash JC 1995 Computer algebra systems: Derive. American Statistician 49(1), 93–99.

Squire W and Trapp G 1998 Using complex variables to estimate derivatives of real functions. SIAM Review 40(1), 110–112.

11

Bounds constraints

Bounds, sometimes called box constraints, are among the most common constraints that users wish to impose on nonlinear optimization or modeling parameters. Fortunately, they are also quite easy to include, although as always care is needed in their implementation and use.

In this treatment, I will generally use the term "bounds." In fact, we will not always impose a lower and upper bounds on every parameter, so we will not always have a "box," and some methods do not lend themselves to imposing a true n-dimensional box in which we will seek a solution.

The general conditions we wish to satisfy are, therefore,

$$\text{lower}_i \leq \text{par}_i \ \text{ and/or } \ \text{par}_i \leq \text{upper}_i \tag{11.1}$$

11.1 Single bound: use of a logarithmic transformation

Suppose we want one of our optimization parameters – call it y – to be nonnegative in our modeling or objective function $m(..., y, ...)$. For the moment, we will ignore other inputs to $m()$, and we will leave the generalization to bounds other than zero until later. Clearly, we could rewrite our function to use a different parameter x and then compute $m(..., x^2, ...)$ or $m(..., \exp(x), ...)$. The second form has, in my experience, worked a little better. It avoids the degeneracy that $-x$ and x give the same value of y, which may give some issues for different optimization tools.

We therefore have the transformation from our original parameter to the internal parameter over which we optimize as

$$x = \log(y) \tag{11.2}$$

Nonlinear Parameter Optimization Using R Tools, First Edition. John C. Nash.
© 2014 John Wiley & Sons, Ltd. Published 2014 by John Wiley & Sons, Ltd.
Companion Website: www.wiley.com/go/nonlinear_parameter

and the reverse as

$$y = \exp(x) \tag{11.3}$$

We need to ensure that we apply these transformations as appropriate, namely,

- At the call to the optimization or parameter determination routine, we need to transform the starting value of y to the equivalent x for each relevant parameter.

- On each call to our objective function or its derivatives, the y must be computed from x. Note that this may be needed for each of several parameters.

- On reporting or using the results, we need to generate the "solution" value of y in the same way.

As an example, let us return to finding the minimum of the Lennard–Jones potential (Section 5.1),

```
# The L-J potential function
ljfne <- function(logr, ep = 0.6501696, sig = 0.3165555) {
    r <- exp(logr)
    fn <- 4 * ep * ((sig/r)^12 - (sig/r)^6)
}
# Use next line to see a graph of the transformed function
# curve(ljfne, from=log(0.3), to=log(1))
min <- optimize(ljfne, interval = c(-10, 10))
print(min)

## $minimum
## [1] -1.0347
##
## $objective
## [1] -0.65017

cat("In the original scale, minimum is at ", exp(min$minimum), "\n")

## In the original scale, minimum is at  0.35532
```

We note that this answer corresponds to that found in Section 5.1, but we have not had to be so careful in specifying the interval in which a minimum is sought.

11.2 Interval bounds: Use of a hyperbolic transformation

A similar idea can be applied when we have open interval bounds, that is,

$$\text{lower}_i < \text{par}_i < \text{upper}_i \tag{11.4}$$

where we do not include the bound in the feasible set. Where all parameters are bounded, we have a "box," but the surface of the box is outside the feasible domain.

This is used in package **dfoptim**, Tim Kelley's variant of the Nelder–Mead polytope algorithm.

The idea is that the range of the hyperbolic tangent function tanh() is $(-1, 1)$. Once again, let us call our parameter y and seek a new, internal, parameter x that is unconstrained while y can only take values in the interval [lower, upper]. This can be accomplished if y and x are related by the expression

$$y = \text{lower} + 0.5(\text{upper} - \text{lower})(1 + \tanh(x)) \qquad (11.5)$$

To reverse the transformation, for example, to get starting parameter values, we use

$$x = a\tanh(2 * (y - \text{lower})/(\text{upper} - \text{lower}) - 1)) \qquad (11.6)$$

Sometimes the tanh() function is too close to a step function or else too slowly changing. In these situations, we need to scale the new parameter x by a number smaller or larger than 1, respectively. In the following example, we use 0.25 to "stretch" the variation of y with x.

11.2.1   Example of the tanh transformation

Let us impose the bounds

$$-4 <= y <= 10 \qquad (11.7)$$

and look at the relationship of y and x. The solid line is the transformation above, while the dotted line uses $\tan h(0.25 * x)$ inside the expression (Figure 11.1).

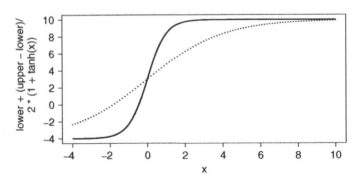

Figure 11.1

11.2.2   A fly in the ointment

The approach we have just presented is quite effective, and as we have indicated, it is used in the function nmkb() in package **dfoptim**. There is, however, one shortcoming, in that we cannot provide any initial parameter **on** a bound. Indeed, we must provide starting values strictly "inside the box." While, in principle, this is not a great difficulty, it can be a nuisance and it needs to be kept in mind when using nmkb, especially inside wrappers such as optimx() where the returned "solutions" will have indicator values such as a very large number for the objective function. An alternative, which I prefer but demands quite a bit of extra programming work, is to check if starting parameters are on a bound.

As an example, consider the (contrived) function below, where we set the lower bounds at 2 for all parameters and the upper bounds at 4 and start at the lower bounds. Here we choose to use a four-parameter example, but the problem is of variable size. The correct solution has only the penultimate parameter free, that is, not on a bound, with the last parameter at the upper bound and the rest at the lower bound.

```
require(optimx)
flb <- function(x) {
    p <- length(x)
    sum(c(1, rep(4, p - 1)) * (x - c(1, x[-p])^2)^2)
}
start <- rep(2, 4)
lo <- rep(2, 4)
up <- rep(4, 4)
ans4a <- try(optimx(start, fn = flb, method = c("nmkb"), lower = lo,
    upper = up))

## Warning: nmkb() cannot be started if any parameter on a bound
```

11.3   Setting the objective large when bounds are violated

A rather old trick in optimization is to set the objective to a very large value whenever a constraint is violated. This will clearly disrupt gradient methods, because we are imposing discontinuities on the objective function. However, it is a technique that can be used for methods that purportedly do not use function values *per se*. The Hooke and Jeeves (1961) method and other pattern search techniques are of this type, but to my knowledge, only hjk() from **dfoptim** of this family is available for R at the time of writing. However, hjkb() from the same package already deals with bounds much more efficiently, because it works with axis search directions.

Some workers argue that the "large value" approach is suitable for use with the Nelder–Mead method, but I tend to agree with John Dennis that it is motivated by ideas of smooth functions, even though it does not explicitly require them. A simple example is illustrative.

```
flbp <- function(x, flo, fup) {
    p <- length(x)
    val <- sum(c(1, rep(4, p - 1)) * (x - c(1, x[-p])^2)^2)
    if (any(x < flo) || any(x > fup))
        val <- 1e+300
    val
}
```

```
start <- rep(2, 4)
lo <- rep(2, 4)
up <- rep(4, 4)
ans4p <- optimx(start, fn = flbp, method = c("hjkb", "Nelder-Mead",
    "nmkb"), flo = lo, fup = up)

## Warning: bounds can only be used with method L-BFGS-B (or Brent)

print(summary(ans4p))

##              p1     p2     p3     p4      value fevals
## hjkb          2 2.0000 2.1091 4.0000 3.2106e+01    332
## Nelder-Mead  NA     NA     NA     NA 8.9885e+307     NA
## nmkb          2 2.0773 2.0000 3.6106 3.7836e+01    499
##            gevals niter convcode  kkt1  kkt2 xtimes
## hjkb           NA    19        0 FALSE FALSE  0.008
## Nelder-Mead    NA    NA     9999    NA    NA  0.004
## nmkb           NA    NA        0 FALSE FALSE  0.060
```

Despite the output, and the use of the name "hjkb" when in fact the unconstrained Hooke and Jeeves method is in use, this is the only method where the "large value" approach has found success. Given this example of a simple function, I do not recommend the approach.

11.4   An active set approach

The approach I have used to bounds in various packages I have put together involves an active set indicator to bounds constraints. The indicator vector maintains a record that tells whether a parameter is at a bound or not. We can use the same index to impose **masks**, that is, fixed parameters.

The particular choice of the set of values for the indicator can be important to make implementation easier. Also it is useful to have a consistent system for the indicator across different packages so code can be reused. The choices I have made are consistent with those used in the BASIC codes I developed with Mary Walker-Smith in (1987). These are as follows.

- Free parameters that are currently NOT at a bound have indicator 1.

- Fixed (masked) parameters are assigned an indicator 0.

- Parameters at an upper bound have indicator -1.

- Parameters at an lower bound have indicator -3.

The choice of -1 and -3 was chosen so adding 2 to the indicator gives either a 1 or -1 value that can be used in a simple multiplication with the gradient to decide if the objective is increasing or decreasing in the direction of the bound. This is useful in determining the gradient projection during the development of a search direction. Again for historical reasons, I call the indicator bdmsk.

The main steps in using this approach repeat for each iteration of a method.

- At the start of the iteration, the gradient (or a proxy) is computed. For each component i of the gradient, if gradient$_i$ * (bdmsk$_i$ + 2) < 0, then the objective DECREASES as we hit the bound, so we cannot proceed in that direction without moving into the infeasible region. In this case, we set gradient$_i$ = 0. Otherwise, we "free" the parameter by setting bdmsk = 1.

- Once we have a search direction t, we want to perform a line search along it. However, we cannot proceed further than the upper or lower bound in each parameter direction. If the current parameters are par, then for $t_i > 0$, we can only go as far as (upper$_i$ − par$_i$)/t_i, while if $t_i < 0$, the limit is (lower$_i$ − par$_i$)/t_i. Clearly, if $t_i = 0$, we need not limit the stepsize. Looping over all components of t, we reduce the steplength allowed for the line search to the smallest value found.

- Once our line search selects a steplength and new set of parameters, we check the bounds and reset indicators appropriately. Clearly, we only need to look at currently free parameters. However, there is the consideration that we may have move to a point very close to a bound. This forces us to use a tolerance for the "closeness." I check if

$$(par_i - lower_i) < ceps * (abs(lower_i) + 1) \tag{11.8}$$

with a similar test on upper bounds. This test is scaled to the size of the bound, with 1 added to avoid difficulties with the very commonly encountered bound at zero. For ceps, I have been using 100 * macheps, where macheps is the floating-point machine precision in use. However, I am always concerned that tolerances such as this will be inappropriate for some problems.

Rvmmin() from the package of the same name uses this approach. I believe, but have not closely examined the codes, that method L-BFGS-B from the function optim() and bobyqa() from package **minqa** use similar techniques.

```
ans4 <- optimx(start, fn = flb, method = c("Rvmmin", "L-BFGS-B",
    "bobyqa"), lower = lo, upper = up, control = list(usenumDeriv = TRUE))

## Warning: Replacing NULL gr with 'numDeriv' approximation

print(summary(ans4))

##           p1 p2      p3 p4  value fevals gevals niter
## Rvmmin     2  2 2.1091  4 32.106    106     22    NA
## L-BFGS-B   2  2 2.1091  4 32.106      7      7    NA
## bobyqa     2  2 2.1091  4 32.106     44     NA    NA
##          convcode  kkt1 kkt2 xtimes
## Rvmmin          0 FALSE TRUE  0.012
## L-BFGS-B        0 FALSE TRUE  0.004
## bobyqa          0 FALSE TRUE  0.000
```

11.5   Checking bounds

Setting bounds, as we shall discuss in the following text, can assist in getting appropriate solutions to optimization and nonlinear equations problems. However, there is no guarantee that software will respect these bounds. It is, therefore, especially important to check that starting values of parameters are consistent with these bounds.

I regard starting values as *infeasible* if they are violate constraints. Some authors, quite reasonably, choose to reset starting values to the nearest bound that makes the starting values feasible. (This will not work for programs such as nmkb that use the transfinite function approach to bounds.)

It is also sensible to check solutions AFTER optimization to ensure that the constraints are still satisfied. I have seen several cases where optimization codes failed to respect constraints, even though that was a stated capability. I have also seen bounds-constrained optimizations that simply failed to achieve the true bounded optimum.

```
require(optextras)
## extract first solution from ans4
ans41 <- ans4[1, ]
c41 <- coef(ans41) ## get coefficients
bdc <- bmchk(c41, lower = lo, upper = up, trace = 1)

## admissible =  TRUE
## maskadded =  FALSE
## parchanged =  FALSE
## At least one parameter is on a bound

print(bdc$bchar)

## [1] "L" "L" "F" "U"
```

11.6   The importance of using bounds intelligently

Why should we bother with bounds?

First, there are problems where the optimal point is actually on one or more bounds. For such problems, ignoring the bounds clearly implies that we have not solved the problem we have been given.

Second, bounds – generally quite "loose" bounds – prevent our optimization tools from jumping far away from the solution due to some artifact of the calculation procedure. In this usage, bounds serve more or less like starting values, except that we now localize the solution to be within a box. My experience is that this saves quite a lot of grief when solving problems where the solutions are generally familiar and there are good reasons to expect solutions to be within some region of the parameter space. That does not mean that we know the answer before we start. Bounds used in this context can still be two or three orders of magnitude apart, so there is still an optimal set of parameters to be found. However, knowing a parameter is between 1 and 1500 is extremely helpful compared to the possibility it is anywhere on the real line.

11.6.1   Difficulties in applying bounds constraints

The first issue when applying bounds constraints is that there are fewer methods for carrying out the optimization. Personally, I feel that all unconstrained optimization tools should be augmented to have a bounds-constrained sibling, because it is fairly straightforward to do this. Nevertheless, at the time of writing, there are a number of tools that are only applicable to unconstrained problems.

A trickier situation arises when gradients or other derivatives are evaluated by numerical approximation. The current working point – that is set of parameters – may be inside the feasible region. However, the typical numerical approximation (see Chapter 10) requires that each parameter be augmented by a small amount and the function evaluated at this new point, which may be OUTSIDE the constraint. This becomes a serious matter if the constraint has been imposed to avoid a computational disaster such as zero-divide, log(0), or sqrt(−1).

It is clearly possible to work out how one might check that a parameter is "close enough" to a bound (or other constraint) that numerical derivative approximation would put a parameter into infeasible territory. However, I have not seen software that actually does this, which while eminently possible, is tedious and imposes potentially large inefficiencies when it is unnecessary.

11.7   Post-solution information for bounded problems

Solutions to bounds-constrained optimization problems pose a reporting issue for statistical interpretation. If a parameter is at a bound, then our interpretation of standard errors is awkward. We generally consider the standard error as a measure of the uncertainty of the quantity reported. However, it is conventional to take such measures as symmetric about the estimated value. The presence of the bound at best renders the interpretation as valid on the half-interval.

There are two computational choices.

- Report the Hessian, gradient, standard errors, and consequent statistics as if there were no bound. This is the choice of the function `optimHess()`.

- Zero the appropriate elements of the Hessian, gradient, and resultant statistics. This is the choice with **nlmrt** and other codes with which I have been part of the development team.

The choice I have made is largely because the optimization method actually zeros the gradient and so on as discussed earlier for working with bounds. The unconstrained Jacobian or Hessian can, of course, be separately computed. I believe both choices have their place. Unfortunately, there does not seem to be much well-structured discussion of the issues of reporting uncertainty in the presence of bounds. Here is a small example. Note that the gradient and Jacobian singular values, while displayed with particular parameters, do not pertain directly to those parameters.

```
require(nlmrt)
y <- c(5.308, 7.24, 9.638, 12.866, 17.069, 23.192, 31.443, 38.558,
    50.156, 62.948, 75.995, 91.972)
t <- 1:12
wdata <- data.frame(y = y, t = t)
anlxb <- nlxb(y ~ 100 * asym/(1 + 10 * strt * exp(-0.1 * rate * t)),
    start = c(asym = 2, strt = 2, rate = 2), lower = c(2, 2, 2),
    upper = c(4, 4, 4), data = wdata, trace = FALSE)
print(anlxb)
```

```
## nlmrt class object: x
## residual sumsquares =   27.72  on  12 observations
##      after  9    Jacobian and  10 function evaluations
## name        coeff       SE      tstat       pval      gradient      JSingval
## asym        2L          NA      NA          NA               0      102.7
## strt        4U          NA      NA          NA               0          0
## rate        2.90391     NA      NA          NA      -7.753e-07          0
```

```
# Get unconstrained Jacobian and its singular values First
# create the function (you can display it if you wish)
myformula <- y ~ 100 * asym/(1 + 10 * strt * exp(-0.1 * rate * t))
# The parameter vector values are unimportant; the names are
# needed
mj <- model2jacfun(myformula, pvec = coef(anlxb))
# Evaluate at the solution parameters
mjval <- mj(prm = coef(anlxb), y = y, t = t)
# and display the singular values
print(svd(mjval)$d)
```

```
## [1] 131.4248    7.6714    2.6137
```

```
# Create the sumsquares function
mss <- model2ssfun(myformula, pvec = coef(anlxb))
# Evaluate gradient at the solution
mgrval <- grad(mss, x = coef(anlxb), y = y, t = t)
print(mgrval)
```

```
## [1]   6.6324e+01 -6.0970e+01  8.1152e-06
```

```
# Evaluate the hessian at the solution (x rather than prm!)
require(numDeriv)
mhesssval <- hessian(mss, x = coef(anlxb), y = y, t = t)
print(eigen(mhesssval)$values)
```

```
## [1] 34617.706   151.004     28.294
```

We note that the gradient is NOT near zero – we are on bounds constraints. The code above illustrates that it is fairly easy to get the quantities that we may want to review concerning the solution, although there is the common R annoyance that the same quantities are given different names in different functions, such as start, pvec, prm, and x. And writing this section reminded me that I am one of the guilty in this regard.

Outside of statistics, optimization commonly looks at a different sort of uncertainty, namely, that of the strictness of the constraints. Clearly, we cannot relax a bound that would lead to log(0) or similar impossibility. However, many bounds are matters of policy, such as a floor price on a commodity in some model of an economy. In this latter situation, we can consider the rate of change of the objective function per unit change in the constraint (i.e., bound). This is the concept of a **shadow price** (http://en.wikipedia.org/wiki/Shadow_price). The concept is most commonly associated with linear programming, and I do not know of any R package that provides such information for bounded solutions of nonlinear function minimization.

Appendix 11.A Function `transfinite`

```
##------------------------------------------------------------
## t r a n s f i n i t e . R
##------------------------------------------------------------

transfinite <- function(lower, upper, n = length(lower)) {
    stopifnot(is.numeric(lower), is.numeric(upper))
    if (any(is.na(lower)) || any(is.na(upper)))
        stop("Any 'NA's not allowed in 'lower' or 'upper' bounds.")
    if (length(lower) != length(upper))
        stop("Length of 'lower' and 'upper' bounds must be equal.")
    if (any(lower == upper))
        stop("No component of 'lower' can be equal to the one in 'upper'.")
    if (length(lower) == 1 && n > 1) {
        lower <- rep(lower, n)
        upper <- rep(upper, n)
    } else if (length(lower) != n)
        stop("If 'length(lower)' not equal 'n', then it must be one.")

    low.finite <- is.finite(lower)
    upp.finite <- is.finite(upper)
    c1 <- low.finite & upp.finite # both lower and upper bounds are finite
    c2 <- !(low.finite | upp.finite) # both lower and upper bounds infinite
    c3 <- !(c1 | c2) & low.finite # finite lower bound, infinite upper bound
    c4 <- !(c1 | c2) & upp.finite # finite upper bound, infinite lower bound

    q <- function(x) {
        if (any(x < lower) || any(x > upper))
            return(rep(NA, n))

        qx <- x
        qx[c1] <- atanh(2 * (x[c1] - lower[c1])/(upper[c1] -
            lower[c1]) - 1)
        qx[c3] <- log(x[c3] - lower[c3])
        qx[c4] <- log(upper[c4] - x[c4])
        return(qx)
    }
```

```
qinv <- function(x) {
    qix <- x
    qix[c1] <- lower[c1] + (upper[c1] - lower[c1])/2 * (1 +
        tanh(x[c1]))
    qix[c3] <- lower[c3] + exp(x[c3])
    qix[c4] <- upper[c4] - exp(x[c4])
    return(qix)
}
return(list(q = q, qinv = qinv))
}
```

References

Hooke R and Jeeves TA 1961 "Direct Search" solution of numerical and statistical problems. Journal of the ACM 8(2), 212–229.

Nash JC and Walker-Smith M 1987 Nonlinear Parameter Estimation: An Integrated System in BASIC. Marcel Dekker, New York. See http://www.nashinfo.com/nlpe.htm for an expanded downloadable version.

12

Using masks

Masks refers to the temporary fixing of one or more parameters and optimizing over the rest. This can be very helpful when one parameter is a setting or control that we later wish to optimize. It can also be useful to fix a parameter to which the objective function is particularly sensitive. In this chapter, we explore this idea, noting that some tools allow such options.

12.1 An example

We return to our Hobbs weeds example with maximum likelihood estimation of the three logistic parameters plus the dispersion. That is, given a set of values of the growth of some quantity y, for example, density of weeds, at times t, we wish to maximize the product of terms of the form

$$\text{term}_i = (1/\text{sqrt}(2\pi\text{sigma}^2)) \exp(-\text{res}_i^2/(2\text{sigma}^2)) \tag{12.1}$$

where

$$\text{res}_i = y_i - b_1/(1 + b_2 \exp(-b_3 * t_i)) \tag{12.2}$$

(see Section 1.2)

Our parameters to adjust are again the b_1, b_2, and b_3, which here are unscaled, as well as the standard error sigma.

12.2 Specifying the objective

We first want to set up the residuals. As R is very good at vector operations, we compute res as a vector and avoid the index i. Let us furthermore work with the

---

Nonlinear Parameter Optimization Using R Tools, First Edition. John C. Nash.
© 2014 John Wiley & Sons, Ltd. Published 2014 by John Wiley & Sons, Ltd.
Companion Website: www.wiley.com/go/nonlinear_parameter

log of the parameters, which is one form of scaling. The code to do this is function
`lhobbs.res`.

```
lhobbs.res <- function(xl, y) {
    # log scaled Hobbs weeds problem - residual base parameters
    # on log(x)
    x <- exp(xl)
    if (abs(12 * x[3]) > 50) {
        # check computability
        rbad <- rep(.Machine$double.xmax, length(x))
        return(rbad)
    }
    if (length(x) != 3)
        stop("hobbs.res - parameter vector n!=3")
    t <- 1:length(y)
    res <- x[1]/(1 + x[2] * exp(-x[3] * t)) - y
}
```

There are a few points to note.

- The parameters used to compute the logistic are the exponentiated values of
  the input parameters. This ensures that these parameters are positive without the need for explicit lower bounds at zero that sometimes give rise to
  computational issues.

- There is an explicit check that our third parameter, in its computational form,
  that is, exponentiated, is not too large.

- The residual is expressed as "model–data," noted in Chapter 10, which is not
  the usual choice made today by statisticians. As we square the residuals, the
  result is the same, and the derivatives of the residual then have the same sign
  as the derivatives of the model. A small point but one that I have found helps
  to reduce errors.

The residuals must now be combined into the likelihood function.

```
lhobbs.lik <- function(xaug, y = y0) {
    # likelihood function including sigma
    xl <- xaug[1:3]
    logSigma <- xaug[4]
    sigma2 = exp(2 * logSigma)
    res <- lhobbs.res(xl, y)
    nll <- 0.5 * (length(res) * log(2 * pi * sigma2) + sum(res *
        res)/sigma2)
}
```

This function explicitly negates the logarithm of the product of the likelihoods.
Thus we can **minimize** this function with respect to the four parameters. Note again
that we use the logarithms of the quantities we want as our parameters.

In this problem, it is relatively easy, although rather tedious, to develop expressions for the gradient via the chain rule, and we will do so. First, we need the
Jacobian of the residuals.

```
lhobbs.jac <- function(xl, y) {
    # scaled Hobbs weeds problem - Jacobian
    x <- exp(xl)
    jj <- matrix(0, 12, 3)
    t <- 1:12
    yy <- exp(-x[3] * t)
    zz <- 1/(1 + x[2] * yy)
    jj[t, 1] <- zz * exp(xl[1])
    jj[t, 2] <- -x[1] * zz * zz * yy * exp(xl[2])
    jj[t, 3] <- x[1] * zz * zz * yy * x[2] * t * exp(xl[3])
    return(jj)
}
```

From this, we can compute the gradient of the sum-of-squares function.

```
lhobbs.g <- function(xl, y) {
    # scaled Hobbs weeds problem - gradient
    shj <- lhobbs.jac(xl, y)
    shres <- lhobbs.res(xl, y)
    shg <- as.vector(2 * (shres %*% shj))
    return(shg)
}
```

Finally, we combine all these to get the gradient of the negative log likelihood.

```
lhobbs.lg <- function(xaug, y = y0) {
    # gradient function including sigma
    xl <- xaug[1:3]
    logSigma <- xaug[4]
    sigma2 = exp(2 * logSigma)
    res3 <- lhobbs.res(xl, y)
    n <- length(res3)
    f3 <- crossprod(res3)
    g3 <- 0.5 * lhobbs.g(xl, y)/sigma2
    gg <- c(g3, (n - f3/sigma2))
}
```

While we could set up a separate function of just three parameters and solve the least squares problem as a way to get preliminary estimates for the logistic parameters, with **masks** we can use the full function just created and fix the sigma parameter. Here is the computation. Note how we use the vector bdmsk to fix the fourth parameter, as well as the very crude starting values.

```
y0 <- c(5.308, 7.24, 9.638, 12.866, 17.069, 23.192, 31.443, 38.558,
    50.156, 62.948, 75.995, 91.972)
require(Rvmmin)

## Loading required package: Rvmmin
## Loading required package: optextras
## Loading required package: numDeriv

start <- c(1, 1, 1, 1)
bdmsk <- c(1, 1, 1, 0) # Cat fix parameter 4 at first
## afix4<-Rvmmintry(start, lhobbs.lik, lhobbs.lg, bdmsk=bdmsk,
```

```
## y=y0, control=list(trace=2))
afix4 <- Rvmmin(start, lhobbs.lik, lhobbs.lg, bdmsk = bdmsk,
    y = y0, control = list(trace = 0))
print(afix4)

## $par
## [1] 1 1 1 1
##
## $value
## [1] 1520
##
## $counts
## [1] 5 2
##
## $convergence
## [1] -1
##
## $message
## [1] "Rvmminb appears to have converged"
##
## $bdmsk
## [1] 1 1 1 0
```

Let us now take the parameters from this run and use them on the four-parameter problem with all parameters free, that is, without masks.

```
start2 <- afix4$par
## new start
print(start2)

## [1] 1 1 1 1

bdmsk <- rep(1, 4) ## EXPLICITLY free all parameters
aall <- Rvmmin(start2, lhobbs.lik, lhobbs.lg, bdmsk = bdmsk,
    y = y0)
print(aall)

## $par
## [1]   5.2791   3.8937  -1.1597  -0.7672
##
## $value
## [1] 7.821
##
## $counts
## [1] 393   91
##
## $convergence
## [1] 0
##
## $message
## [1] "Rvmminb appears to have converged"
```

Finally, let us return to the full four-parameter problem and then compare the work done by the two approaches.

```
## [1] 1 1 1 1
## $par
## [1]   5.2791   3.8937 -1.1597 -0.7672
##
## $value
## [1] 7.821
##
## $counts
## [1] 393   91
##
## $convergence
## [1] 0
##
## $message
## [1] "Rvmminb appears to have converged"
##
##
## Comparison of work: Masked       vs.      Free
##      Functions            398             393
##      Gradients             93              91
##
##  Exponentiated parameters, last is sigma:
## [1] 196.1863  49.0916   0.3136   0.4643
```

Here we see that there is some saving in workload, although on modern computers the time saving is not noticeable. More importantly, however, it is sometimes difficult to get a solution directly when all parameters are left free, while the staged approach is able to succeed.

12.3   Masks for nonlinear least squares

Package **nlmrt** allows masks to be specified using the names of the parameters. Here is a small example of the Bates form of the three-parameter logistic where we fix the upper asymptote to a particular value.

```
weeddata <- data.frame(y = y0, t = 1:12)
mystart <- c(Asym = 250, xmid = 6, scal = 1) # This sets the Asym value
require(nlmrt) # Ensure tools available

## Loading required package: nlmrt
maskrun <- nlxb(y ~ Asym/(1 + exp((xmid - t)/scal)), start = mystart,
    data = weeddata, mask = c("Asym"), trace = FALSE)
maskrun

## nlmrt class object: x
## residual sumsquares =  6.0738  on  12 observations
##     after  6    Jacobian and  7 function evaluations
##    name       coeff       SE      tstat     pval      gradient   JSingval
## Asym        250  M         NA       NA       NA              0       52.8
## xmid      13.8144          NA       NA       NA      3.094e-06       12.9
## scal      3.43303          NA       NA       NA     -6.946e-06          0
```

In the above output, we note that standard error estimates and *t*-statistics are omitted. This is because the Jacobian of the reduced dimension problem is now singular. We can, of course, recast the problem, as we do in the following. However, it is part of my "to-do" list to adjust the package functions to provide the appropriate estimates.

```
mystart2 <- c(xmid = 6, scal = 1) # This sets the Asym value
maskrun2 <- nlxb(y ~ 250/(1 + exp((xmid - t)/scal)), start = mystart2,
    data = weeddata)
maskrun2

## nlmrt class object: x
## residual sumsquares =  6.0738  on  12 observations
##     after  6    Jacobian and  7 function evaluations
##    name      coeff       SE      tstat      pval      gradient    JSingval
## xmid       13.8144    0.04945    279.3         0     3.094e-06       52.8
## scal        3.43303   0.03772       91  6.661e-16   -6.946e-06       12.9
```

12.4   Other approaches to masks

Clearly, we can employ the **idea** of masks by explicitly coding the appropriate functions. This is, of course, more work when our goal is to estimate the fully parameterized model. It has surprised the author that the idea of masks is not more common in nonlinear parameter estimation software. One tool that does include the idea is AD Model Builder (ADMD), where it is referred to as **phases**. In ADMB, parameters are specified via lower and upper bounds and an integer giving the "phase" at which they will be allowed to participate in the optimization.

In the **stats4** package that is distributed with base R the mle() function claims to allow the user to specify parameters with fixed values, through I have not found success in trying to use this function. By contrast, the very similar mle2() from **bbmle** (Bolker and Team, 2013) seems to work quite well. It includes the same facility for fixing parameters, that is, fixed=list(parameter= value). There is an example in Section 21.1.

The package **maxLik** (Toomet and Henningsen, 2008) also allows for what the authors call "fixed parameters." Here the parameters to be fixed (masked) are specified in a vector giving the positions of the values to be fixed in the parameter vector. For situations with many parameters, this will be tidier than my use of an indicator of zero in the appropriate position in bdmsk. Even better – an idea I shall likely copy – they allow for a vector of character names of the relevant parameters.

References

Bolker B and Team RDC 2013 *bbmle: Tools for general maximum likelihood estimation*. R package version 1.0.13.

Toomet O and Henningsen A 2008 Sample selection models in r: package sample selection. Journal of Statistical Software 27(7), 1–23.

13

Handling general constraints

In this chapter, we consider constraints that are neither bounds nor masks. If there are many constraints, then the problem is usually considered **mathematical programming**, see Chapter 14, and it is common that the constraints are then often linear. There may also be other features, such as variables that are all integers, or there is a mix of integer and real parameters. Such problems have generally been less common in statistics, but there appears to be growing interest.

In this chapter, however, we consider problems with just a few constraints, so that the main focus of attention is still the objective function.

13.1 Equality constraints

Equality constraints are, in my experience, usually more troublesome than inequality constraints. This might seem paradoxical. After all, each equality constraint should reduce the dimension of the problem by one parameter. The difficulty is in choosing WHICH parameter to eliminate.

Nash (1979, p. 184) presents a problem (with seemingly erroneous results reported) that is apparently a linear regression except that four of the coefficients are related. We have six variables $y_1, \ldots, y_6$ and want a model

$$y_1 = b_1 + b_2 * y_2 + b_3 * y_3 + b_4 * y_4 + b_5 * y_5 + b_6 * y_6 + \text{error} \qquad (13.1)$$

However, there is a constraint so that

$$b_3 * b_6 = b_4 * b_5 \qquad (13.2)$$

In this example, it is very obvious that we can solve for any one of the four parameters. The data for the problem is read from a comma-separated value (CSV) file. From the origin of the data set and problem, I call this the Hassan problem.

Nonlinear Parameter Optimization Using R Tools, First Edition. John C. Nash.
© 2014 John Wiley & Sons, Ltd. Published 2014 by John Wiley & Sons, Ltd.
Companion Website: www.wiley.com/go/nonlinear_parameter

First, let us find the unconstrained solution and check the violation of the constraint.

```
hn <- read.csv("hassan182.csv", header = TRUE)
convi <- function(x, dta) {
    # dta not used
    h <- rep(NA, 1) # vector value
    h[1] <- x[3] * x[6] - x[4] * x[5]
    h
}
uncmod <- lm(one ~ two + three + four + five + six, data = hn)
uncmod

##
## Call:
## lm(formula = one ~ two + three + four + five + six, data = hn)
##
## Coefficients:
## (Intercept)        two        three         four
##     -1.1352     0.9460       0.0997      -0.0276
##        five        six
##    -31.4896    56.1636

bu <- coef(uncmod)
cat("Constraint violation = b3*b6-b4*b5 =", as.numeric(convi(bu)),
    "\n")

## Constraint violation = b3*b6-b4*b5 = 4.7287

cat("Sum of squared residuals = ", as.numeric(crossprod(uncmod$residuals)),
    "\n")

## Sum of squared residuals =  891.99
```

A minor warning. The nonlinear least squares codes nls(), nlxb(), and nlsLM() take model expressions similar to BUT NOT THE SAME AS those used by lm() to build uncmod. As linear models are a special case of nonlinear ones, we could have used these functions to perform the calculations above. To do this, we need to explicitly add our coefficients, namely,

```
uncmod2 <- nls(one ~ b1 + two * b2 + three * b3 + four * b4 +
    five * b5 + six * b6, data = hn)

## Warning: No starting values specified for some parameters.
## Initializing 'b1', 'b2', 'b3', 'b4', 'b5', 'b6' to '1.'.
## Consider specifying 'start' or using a selfStart model

summary(uncmod2)

##
## Formula: one ~ b1 + two * b2 + three * b3 + four * b4 + five * b5 + six *
##    b6
##
## Parameters:
##    Estimate Std. Error t value Pr(>|t|)
## b1  -1.1352    41.3043   -0.03    0.978
```

```
## b2     0.9460     0.1917     4.93     8e-05 ***
## b3     0.0997     0.0576     1.73     0.099 .
## b4    -0.0276     0.0305    -0.91     0.376
## b5   -31.4895    34.9621    -0.90     0.378
## b6    56.1637    37.3244     1.50     0.148
## - - -
## Signif. codes:  0 '***' 0.001 '**' 0.01 '*' 0.05 '.' 0.1 ' ' 1
##
## Residual standard error: 6.68 on 20 degrees of freedom
##
## Number of iterations to convergence: 3
## Achieved convergence tolerance: 4.18e-07
```

There are three approaches to a solution of the constrained problem that we will examine here.

- Solve the constraint equation for one of the parameters in terms of the others and substitute this into the model above.

- Use tools for nonlinear optimization subject to nonlinear equality (or also inequality) constraints, of which **alabama** and **Rsolnp** seem to be the most promising;

- Define a penalized objective function where we add a positive factor times the sum of squares of the constraint violation(s) and solve sequentially, with increasing values of the penalty factor. Parameters from each solution are fed in as starting values for the next solution with stronger penalty.

For the present problem, which has rather nasty properties, we will see that it is very easy to get answers uncomfortably far from the best possible.

13.1.1   Parameter elimination

One difficulty with "solving" for one of the parameters in this problem is that we have the possibility of a zero divide. For example, if we solve for

$$b_3 = b_4 * b_5/b_6 \tag{13.3}$$

and b_6 gets close to zero, we can expect our algorithms and programs to get into trouble. Despite this possibility, we can still try to approach the problem by solving for one parameter in terms of the rest and using nonlinear least squares on the resulting model, or else minimize the sum of squares of these residuals with an optimizer.

```
res3 <- function(x, dta) {
    b1 <- x[1]
    b2 <- x[2]
    # b3<-x[3]
    b4 <- x[3]
    b5 <- x[4]
```

```
    b6 <- x[5]
    res <- dta$two * b2 + dta$three * b4 * b5/b6 + dta$four *
        b4 + dta$five * b5 + dta$six * b6 + b1 - dta$one
    return(-res)
}
ss3 <- function(x, dta) {
    res <- res3(x, dta)
    sum(res * res)
}
require(nlmrt)
```

Loading required package: nlmrt

```
require(optimx)
```

Loading required package: optimx

```
st3 <- setNames(bu[-3], c("b1", "b2", "b4", "b5", "b6"))
anlfb3 <- try(nlfb(start = st3, resfn = res3, dta = hn))
if (class(anlfb3) != "try-error") {
    cnb3 <- coef(anlfb3)
    print(cnb3)
    cat("RSS=", crossprod(anlfb3$resid), "\n")
} else {
    cat("Error - failed\n")
}
```

```
##          b1          b2          b4          b5          b6
## -28.680156    1.036184   -0.042352  -41.310598   79.207696
## attr(,"pkgname")
## [1] "nlmrt"
## RSS= 985.52
```

```
## derived_b3
cnb3["b4"] * cnb3["b5"]/cnb3["b6"]
```

```
##        b4
## 0.022089
```

```
## Try optimx
aop3 <- optimx(st3, fn = ss3, gr = "grnd", dta = hn,
 control = list(all.methods = TRUE))
```

```
## Warning: Replacing NULL gr with 'grnd' approximation
## Loading required package: numDeriv
## Warning: Parameters or bounds appear to have different scalings.
##   This can cause poor performance in optimization.
##   It is important for derivative free methods like BOBYQA, UOBYQA, NEWUOA.
## Warning: bounds can only be used with method L-BFGS-B (or Brent)
## Warning: bounds can only be used with method L-BFGS-B (or Brent)
## Warning: bounds can only be used with method L-BFGS-B (or Brent)
## Warning: Function evaluation limit exceeded -- may not converge.
```

```
summary(aop3, order = value)
```

```
##             b1        b2         b4       b5      b6
## ucminf  -28.6803  1.03618  -0.042352  -41.311  79.208
## Rvmmin  -28.6803  1.03618  -0.042352  -41.311  79.208
```

```
## nlminb       -28.6803 1.03618 -0.042352 -41.311 79.208
## Rcgmin       -28.6807 1.03619 -0.042352 -41.310 79.207
## nmkb         -28.6803 1.03618 -0.042354 -41.308 79.209
## BFGS         -25.3920 1.01751 -0.043002 -35.760 82.639
## CG           -25.3920 1.01751 -0.043002 -35.760 82.639
## Nelder-Mead  -25.3920 1.01751 -0.043002 -35.760 82.639
## L-BFGS-B     -25.3920 1.01751 -0.043002 -35.760 82.639
## hjkb         -13.2112 0.96400 -0.031632 -39.492 71.582
## bobyqa        -1.1530 0.91213 -0.018279 -31.496 57.114
## newuoa        -1.1312 0.91182 -0.018244 -31.399 57.140
## spg           -1.1713 0.91260 -0.017788 -31.549 56.304
## nlm           -1.1354 0.91246 -0.017685 -31.490 56.164
##                 value fevals gevals niter convcode  kkt1
## ucminf         985.52     27     27    NA        0  TRUE
## Rvmmin         985.52     68     23    NA        0  TRUE
## nlminb         985.52     45     24    23        0  TRUE
## Rcgmin         985.52   1228    310    NA        1  TRUE
## nmkb           985.52   1050     NA    NA        0 FALSE
## BFGS           986.67     51     51    NA        0 FALSE
## CG             986.67     51     51    NA        0 FALSE
## Nelder-Mead    986.67     51     51    NA        0 FALSE
## L-BFGS-B       986.67     51     51    NA        0 FALSE
## hjkb           995.02  10007     NA    16        1 FALSE
## bobyqa        1039.36    235     NA    NA        0 FALSE
## newuoa        1039.43    217     NA    NA        0 FALSE
## spg           1041.92   2764     NA  1501        1 FALSE
## nlm           1042.51     NA     NA     9        0 FALSE
##              kkt2 xtimes
## ucminf      FALSE  0.080
## Rvmmin      FALSE  0.068
## nlminb      FALSE  0.064
## Rcgmin      FALSE  0.888
## nmkb        FALSE  0.116
## BFGS        FALSE  0.132
## CG          FALSE  0.136
## Nelder-Mead FALSE  0.136
## L-BFGS-B    FALSE  0.132
## hjkb        FALSE  0.465
## bobyqa      FALSE  0.012
## newuoa      FALSE  0.008
## spg         FALSE  4.280
## nlm         FALSE  0.000
```

The reader may rightly be dismayed by the variety of results with **optimx**. However, there is general agreement between the most successful methods (those with the smallest sum of squared residuals) and the nonlinear least squares solution.

13.1.2   Which parameter to eliminate?

We chose to eliminate b_3 in the computations just completed. Is there any reason to choose this over any of b_4, b_5, or b_6? As it turns out, the choice can be important. In particular, when b_6 is eliminated, the best nonlinear least squares solution by nlfb() or **optimx** has a residual sum of squares of 1008.334. This is clearly not close to the optimal value of 985.5175 found when any of the other parameters is eliminated. Trying a starting vector of all 1's does allow two of the **optimx** methods to find the lower value, but nlfb() still ends up at the higher sum of squares. The unconstrained parameter b_6 is the largest numerically, and in the elimination of b_3, it appears as a divisor. When we eliminate b_6, the rather small b_3 is our divisor, and unfortunate steps during optimization could risk a zero divide. However, this issue does not seem to affect the elimination of b_5, which is also quite large in magnitude.

13.1.3   Scaling and centering?

In the process of developing this example, I explored the conventional advice of centering the variables, that is, subtracting their means. For linear regression, this often stabilizes the computations and reduces the dimensionality of the normal equations (or their equivalent) by 1. One can also divide by the standard deviation to standardize the variables.

In the present example, however, centering or standardizing the variables introduces a change that complicates the constraint. Unless such stabilization offers a very large advantage, I would avoid it. Simple scalings by powers of 10 are, however, useful without the necessity of algebraic manipulations. I found that all four choices of parameter to eliminate will give at least one satisfactory solution with **optimx** if I scale the columns of the data by multiplying them respectively with

$$1e + 00 \quad 1e - 02 \quad 1e - 01 \quad 1e - 03 \quad 1e + 02 \quad 1e + 00$$

However, elimination of b_6 still failed to get a nonlinear least squares solution.

13.1.4   Nonlinear programming packages

Let us try using the constrOptim.nl() and auglag() augmented Lagrangian methods of package **alabama** and the solnp() function of **Rsolnp**. All these handle inequality constraints too, but for now, we will use just the equality constraint capabilities. Once again, we must caution that the calling syntax of different functions is not standardized, and in particular, note the eqB field for solnp(). We include the Jacobian calculation in the code but do not necessarily use it. Note that there is one other package, **Rdonlp2**, which is an interface to Peter Spelluci's DONLP2 nonlinear programming code. However, there are licensing restrictions on this R-forge package, for which reason I am reluctant to pursue its use here. Here is the setup for the other codes.

```
resall <- function(x, dta) {
    b1 <- x[1]
    b2 <- x[2]
    b3 <- x[3]
    b4 <- x[4]
    b5 <- x[5]
    b6 <- x[6]
```

```
    res <- dta$two * b2 + dta$three * b3 + dta$four * b4 + dta$five *
        b5 + dta$six * b6 + b1 - dta$one
    return(-res)
}
jacall <- function(x, dta) {
    b1 <- x[1]
    b2 <- x[2]
    b3 <- x[3]
    b4 <- x[4]
    b5 <- x[5]
    b6 <- x[6]
    nobs <- dim(dta)[1]
    JJ <- cbind(rep(-1, nobs), -dta$two, -dta$three, -dta$four,
        -dta$five, -dta$six)
}
ssall <- function(x, dta) {
    res <- resall(x, dta)
    sum(res * res)
}

stones <- rep(1, 6) # trivial start
names(stones) <- c("b1", "b2", "b3", "b4", "b5", "b6")
stunc <- coef(uncmod) # unconstrained parameter start
```

We will not present all the output, which is quite voluminous. The calling sequences for constrOptim.nl(), auglag(), and solnp() are as follows for start stones.

```
require(alabama)
aalabco <- constrOptim.nl(par = stones, fn = ssall, heq = convi,
    dta = hn)
aauglag <- auglag(stones, fn = ssall, heq = convi, dta = hn)
require(Rsolnp)
asolnp <- solnp(stones, fun = ssall, eqfun = convi, eqB = c(0),
    dta = hn)
```

```
## Loading required package: alabama

## Start from all ones -- constrOptim.nl
##          b1         b2         b3         b4         b5
## -21.673745   1.005383   0.016668  -0.051264 -30.666798
##          b6
##   94.321144
## constrOptim.nl: sumsquares = 1003.8    constr. violation= 7.8447e-08

## Start from unconstrained model -- constrOptim.nl
## (Intercept)        two      three        four        five
## -16.404246   0.979924   0.018891  -0.043194 -37.585187
##         six
##   85.940314
## constrOptim.nl: sumsquares = 997.18    constr. violation= 9.2954e-08
```

```
## Start from all ones -- auglag
##         b1          b2          b3          b4          b5
## -13.684176    0.968233    0.014537   -0.045940  -28.667509
##         b6
##   90.596584
## auglag: sumsquares = 1005.3    constr. violation= 1.9809e-11


## Start from unconstrained model -- auglag
## (Intercept)         two       three        four        five
##  -18.188436    0.982143    0.018380   -0.042621  -37.150893
##         six
##    86.147218
## auglag: sumsquares = 993.04    constr. violation= 1.3769e-08


## Loading required package: Rsolnp
## Loading required package: truncnorm
## Loading required package: parallel

## Start from all ones -- solnp
##         b1          b2          b3          b4          b5
##  15.3698499   0.9193906   0.1548982   0.0078678 -55.1476935
##         b6
##  -2.8011496
## solnp: sumsquares = 1008.3    constr. violation= -8.043e-13
## Start from unconstrained model -- solnp
## (Intercept)         two       three        four        five
##  -28.728953    1.036451    0.022111   -0.042383  -41.326619
##         six
##   79.215658
## solnp: sumsquares = 985.52    constr. violation= 2.5975e-12
```

Again the results are not consistent. Note that `auglag()` can take starting parameters that are not feasible, while `constrOptim.nl()` requires a start that satisfies the constraint(s).

13.1.5   Sequential application of an increasing penalty

Penalty functions are another way to try to impose constraints, albeit approximately. We construct an objective function from the original one and then add some weighted multiple of a positive function of the constraint violation. In our case, the constraint is an equality, so we square its value. We scale this with the factor `pen` that has a default value of 100 but which is an argument to our following modified objective function.

To automate the process, we start with `pen=10` and increase it successively until the residual sum of squares does not change iteration to iteration. This is NOT a guaranteed procedure, of course.

```
hassanp.f <- function(x, pen = 100, dta) {
    mod <- x[1] + x[2] * dta$two + x[3] * dta$three + x[4] *
        dta$four + x[5] * dta$five + x[6] * dta$six
```

```
    res <- mod - dta$one
    pfn <- x[3] * x[6] - x[4] * x[5]
    fun <- sum(res * res) + pen * pfn * pfn # return(fun)
}
xx <- coef(uncmod)
bestRSS <- 1e+300
rss <- 1e+299 # make it different from bestRSS to start loop
pen <- 10
while (rss != bestRSS) {
    bestRSS <- rss
    pen = 10 * pen
    anmp <- optim(par = xx, fn = hassanp.f, dta = hn, pen = pen)
    xx <- anmp$par # for next iteration
    rss <- ssall(xx, dta = hn)
    cat("RSS = ", rss, " for pen=", pen, "\n")
    cat("     Constraint violation =", convi(xx, dta = hn), "\n")
}
```

```
## RSS =  985.43   for pen= 100
##      Constraint violation = 0.13551
## RSS =  985.54   for pen= 1000
##      Constraint violation = 0.0142
## RSS =  985.54   for pen= 10000
##      Constraint violation = 0.00050568
## RSS =  985.52   for pen= 1e+05
##      Constraint violation = 0.00016767
## RSS =  985.52   for pen= 1e+06
##      Constraint violation = 1.3581e-05
## RSS =  985.52   for pen= 1e+07
##      Constraint violation = 1.6515e-06
## RSS =  985.52   for pen= 1e+08
##      Constraint violation = 4.3183e-07
## RSS =  985.52   for pen= 1e+09
##      Constraint violation = 4.1873e-08
## RSS =  985.52   for pen= 1e+10
##      Constraint violation = 1.7881e-08
## RSS =  985.52   for pen= 1e+11
##      Constraint violation = 3.9851e-09
## RSS =  985.52   for pen= 1e+12
##      Constraint violation = -1.6715e-09
## RSS =  985.52   for pen= 1e+13
##      Constraint violation = 4.4855e-10
## RSS =  985.52   for pen= 1e+14
##      Constraint violation = 1.375e-11
## RSS =  985.52   for pen= 1e+15
##      Constraint violation = -2.5066e-11
## RSS =  985.52   for pen= 1e+16
##      Constraint violation = 1.2307e-11
## RSS =  985.52   for pen= 1e+17
##      Constraint violation = -3.7594e-12
## RSS =  985.52   for pen= 1e+18
##      Constraint violation = 4.4031e-13
## RSS =  985.52   for pen= 1e+19
##      Constraint violation = -1.8341e-13
## RSS =  985.52   for pen= 1e+20
##      Constraint violation = 1.5277e-13
## RSS =  985.52   for pen= 1e+21
##      Constraint violation = -2.2871e-14
```

```
## RSS =   985.52   for pen= 1e+22
##     Constraint violation = -5.5511e-15
## RSS =   985.52   for pen= 1e+23
##     Constraint violation = 8.8818e-16
## RSS =   985.52   for pen= 1e+24
##     Constraint violation = 8.8818e-16

## Best parameters
print(xx)

## (Intercept)        two       three        four        five
##  -28.564318   1.035479    0.021894   -0.042405  -40.985170
##         six
##   79.382823
```

Note that there is a sensitivity to the initial penalty and the factor by which it is increased, so that it is possible to converge to a solution with the objective above 1000 (possibly a plateau or a local minimum, because similar proposed answers are seen elsewhere).

Penalty methods like this are sometimes helpful to get a quick handle on a problem. Note that we used the default Nelder–Mead optimizer to do the work. This approach would, therefore, make sense if we were blocked from installing software that is more suited to the problem, for example, if we were at a location without network service.

13.2   Sumscale problems

A particular class of optimization problem that arises quite often is one that may be called a **sumscale** problem. Here we wish to maximize or minimize an objective function $f(\mathbf{x})$ has parameters that are constrained by some scaling, so that $q(\mathbf{x}) =$ constant, where this function $q()$ involves a sum of the parameters, their squares, or similar simple function.

The following are the examples of this type of objective function.

- The maximum volume of a regular polyhedron where the sum of the lengths of the sides is fixed.

- The minimum negative log likelihood for a multinomial model.

- The Rayleigh quotient for the maximal or minimal eigensolutions of a matrix, where the eigenvectors should be normalized so the square norm of the vector is 1.

For the moment, let us consider a basic example, which is

$$\textbf{Problem MML: Minimize } \left(- \prod \mathbf{x} \right) \text{ subject to } \sum \mathbf{x} = 1$$

This is a very simplified version of the multinomial maximum likelihood problem, suggested by Gabor Grothendieck as a test problem of this type. One approach to solving this is to eliminate the last parameter and optimize over the other $n - 1$ where n is the number of parameters. Let us try it for a few values of n. We turn off some tests and warnings to keep the output tidy.

```
require(optimx, quietly = TRUE)
# load a range of optimization methods
pr <- function(y) {
    -prod(y) * (1 - sum(y))
}
pr.g <- function(x) {
    g <- -prod(x) * (1 - sum(x))/x + prod(x)
}
n <- 2
cat("  n", "\t", "  x1", "\t\t\t", " 1e5*(x1-1/n)\n")
options(digits = 6)
while (n <= 6) {
    st <- 1:(n - 1)/(n * n)
    ctrl <- list(starttests = FALSE, trace = 0, dowarn = FALSE)
    ans <- optimx(st, pr, pr.g, method = "Rvmmin", control = ctrl)
    cc <- coef(ans)
    cat(n, "\t", (1/n), "\n")
    print(cc)
    n <- n + 1
}
```

```
##    n      x1        1e5*(x1-1/n)
## 2    0.5
##        [,1]
## [1,] 0.35
## 3    0.333333
##                  p1          p2
## Rvmmin 0.333333 0.333333
## 4    0.25
##            p1     p2     p3
## Rvmmin 0.0625 0.125 0.1875
## 5    0.2
##          p1    p2    p3    p4
## Rvmmin 0.04 0.08 0.12 0.16
## 6    0.166667
##              p1         p2        p3        p4        p5
## Rvmmin 0.0277778 0.0555556 0.0833333 0.111111 0.138889
```

The "solution" is, as we expect, simply $x_j = 1/n$ for all j. For very small n, this approach works fine. However, we see that very soon there are computational problems as the size of the problem increases. As the sum of the parameters is constrained to be equal to 1, the parameters are of the order of $1/n$, so the function is therefore of the order of $1/(n^n)$, which underflows around $n = 144$ in R. In reality, the rot sets in much sooner.

We can do much better if we work with the log of the objective function. The following calculation handles very large numbers of parameters, and we only display the analytic answer along with the first five parameter estimates. Note that we change to using **Rcgmin** which uses a method adapted for large numbers of parameters.

```
nll <- function(y) {
    if ((any(y <= 10 * .Machine$double.xmin)) || (sum(y) > 1 -
        .Machine$double.eps))
        .Machine$double.xmax else -sum(log(y)) - log(1 - sum(y))
}
nll.g <- function(y) {
    -1/y + 1/(1 - sum(y))
} # so far not safeguarded
n <- 1000
cat(" n", "\t", " x1", "\t\t\t", " 1e5*(x1-1/n)\n")

## n       x1        1e5*(x1-1/n)

options(digits = 6)
while (n <= 10000) {
    st <- 1:(n - 1)/(n * n)
    ctrl <- list(starttests = FALSE, trace = 0, dowarn = FALSE)
    ans <- optimx(st, nll, nll.g, method = "Rcgmin", control = ctrl)
    cc <- coef(ans)
    cat(n, "\t", (1/n), "\n")
    print(cc[1:5])
    n <- n + 1000
}

## 1000    0.001
## [1] 0.001 0.001 0.001 0.001 0.001
## 2000    5e-04
## [1] 0.000500000 0.000500000 0.000499999 0.000500000
## [5] 0.000500000
## 3000    0.000333333
## [1] 0.000333333 0.000333334 0.000333333 0.000333333
## [5] 0.000333333
## 4000    0.00025
## [1] 0.000250002 0.000250000 0.000250000 0.000250000
## [5] 0.000250000
## 5000    2e-04
## [1] 0.009956240 0.000244116 0.000446017 0.000329520
## [5] 0.000181000
## 6000    0.000166667
## [1] 0.03565142 0.01697512 0.01075478 0.00764869 0.00578853
## 7000    0.000142857
## [1] 0.04500462 0.01959146 0.01113446 0.00691942 0.00440483
## 8000    0.000125
## [1] 0.00916996 0.00419875 0.00254975 0.00173191 0.00124712
## 9000    0.000111111
## [1] 0.000111111 0.000111111 0.000111111 0.000111111
## [5] 0.000111111
## 10000   1e-04
## [1] 0.012336191 0.004579923 0.002023620 0.000785113
## [5] 0.000124854
```

We see that we are doing quite well up to $n = 5000$ but thereafter get uneven results. The $n = 9000$ result is surprisingly good, but optimization methods do occasionally take a lucky step. They can equally be "unlucky." Consider the $n = 5$ case with several optimizers, where we first do not specify the gradient, then we give the analytic gradient, and finally the gradient and bounds on the parameters.

```
require(optimx, quietly = TRUE)
n <- 5
mset <- c("L-BFGS-B", "BFGS", "CG", "spg", "ucminf", "nlm", "nlminb",
    "Rvmmin", "Rcgmin")
a5 <- optimx(2:n/n^2, nll, method = mset, control
= list(dowarn = FALSE))
```

```
## Warning: bounds can only be used with method L-BFGS-B (or Brent)
## Warning: bounds can only be used with method L-BFGS-B (or Brent)
```

```
a5g <- optimx(2:n/n^2, nll, nll.g, method = mset, control = list(dowarn =
FALSE))
```

```
## Warning: bounds can only be used with method L-BFGS-B (or Brent)
## Warning: bounds can only be used with method L-BFGS-B (or Brent)
```

```
a5gb <- optimx(2:n/n^2, nll, nll.g, lower = 0, upper = 1, method = mset,
    control = list(dowarn = FALSE))
summary(a5, order = value)
```

##		p1	p2	p3	p4	value	fevals
##	Rcgmin	0.2	0.200000	0.200000	0.200000	8.04719e+00	51
##	spg	0.2	0.200000	0.200000	0.200000	8.04719e+00	17
##	Rvmmin	0.2	0.200000	0.200000	0.200000	8.04719e+00	22
##	ucminf	0.2	0.200000	0.200000	0.200000	8.04719e+00	14
##	nlm	0.2	0.199999	0.200000	0.200000	8.04719e+00	NA
##	nlminb	0.2	0.199999	0.199999	0.199999	8.04719e+00	23
##	L-BFGS-B	NA	NA	NA	NA	8.98847e+307	NA
##	BFGS	NA	NA	NA	NA	8.98847e+307	NA
##	CG	NA	NA	NA	NA	8.98847e+307	NA
##		gevals	niter	convcode	kkt1	kkt2	xtimes
##	Rcgmin	18	NA	0	NA	NA	0.004
##	spg	NA	13	0	TRUE	TRUE	0.056
##	Rvmmin	12	NA	2	NA	NA	0.004
##	ucminf	14	NA	0	TRUE	TRUE	0.000
##	nlm	NA	11	0	TRUE	TRUE	0.000
##	nlminb	66	12	0	TRUE	TRUE	0.004
##	L-BFGS-B	NA	NA	9999	NA	NA	0.000
##	BFGS	NA	NA	9999	NA	NA	0.000
##	CG	NA	NA	9999	NA	NA	0.004

```
summary(a5g, order = value)
```

##		p1	p2	p3	p4	value	fevals
##	ucminf	0.2	0.200000	0.200000	0.200000	8.04719e+00	14
##	Rvmmin	0.2	0.200000	0.200000	0.200000	8.04719e+00	43
##	spg	0.2	0.200000	0.200000	0.200000	8.04719e+00	17
##	Rcgmin	0.2	0.200000	0.200000	0.200000	8.04719e+00	28
##	nlm	0.2	0.199999	0.200000	0.200000	8.04719e+00	NA
##	nlminb	0.2	0.199999	0.199999	0.199999	8.04719e+00	23
##	L-BFGS-B	NA	NA	NA	NA	8.98847e+307	NA
##	BFGS	NA	NA	NA	NA	8.98847e+307	NA
##	CG	NA	NA	NA	NA	8.98847e+307	NA
##		gevals	niter	convcode	kkt1	kkt2	xtimes
##	ucminf	14	NA	0	TRUE	TRUE	0.000
##	Rvmmin	12	NA	0	TRUE	TRUE	0.004
##	spg	NA	13	0	TRUE	TRUE	0.052
##	Rcgmin	12	NA	0	TRUE	TRUE	0.000
##	nlm	NA	11	0	TRUE	TRUE	0.000

```
## nlminb      12    12       0 TRUE TRUE  0.000
## L-BFGS-B    NA    NA    9999   NA   NA  0.000
## BFGS        NA    NA    9999   NA   NA  0.004
## CG          NA    NA    9999   NA   NA  0.000


summary(a5gb, order = value)

##              p1       p2       p3       p4       value fevals
## Rvmmin      0.2 0.200000 0.200000 0.200000 8.04719e+00     38
## Rcgmin      0.2 0.200000 0.200000 0.200000 8.04719e+00     18
## spg         0.2 0.200000 0.200000 0.200000 8.04719e+00     18
## nlminb      0.2 0.199999 0.199999 0.199999 8.04719e+00     23
## L-BFGS-B     NA       NA       NA       NA 8.98847e+307    NA
##          gevals niter convcode kkt1 kkt2 xtimes
## Rvmmin       14    NA        0 TRUE TRUE  0.004
## Rcgmin       10    NA        0 TRUE TRUE  0.004
## spg          NA    13        0 TRUE TRUE  0.056
## nlminb       12    12        0 TRUE TRUE  0.000
## L-BFGS-B     NA    NA     9999   NA   NA  0.000
```

The particular L-BFGS-B failure is due to the optimization method trying to compute the gradient where sum(x) is greater than 1, although why a set of parameters like that has been generated has not been investigated. It seems likely the failure of **Rvmmin** in the bounded case has a similar origin.

13.2.1   Using a projection

Objective functions defined by $(-1) * \prod \mathbf{x}$ or $(-1) * \sum \log(\mathbf{x})$ will change with the scale of the parameters. Moreover, the constraint $\sum \mathbf{x} = 1$ effectively imposes the scaling $\mathbf{x}_{scaled} = \mathbf{x} / \sum \mathbf{x}$. The optimizer spg() from package **BB** allows us to project our search direction to satisfy constraints. Thus, we could use the following approach.

```
require(BB, quietly = TRUE)
nllrv <- function(x) {
    -sum(log(x))
}
nllrv.g <- function(x) {
    -1/x
}
proj <- function(x) {
    x/sum(x)
}
n <- 5
aspg <- spg(par = (1:n)/n^2, fn = nllrv, gr = nllrv.g, project = proj)

## iter:  0  f-value:  11.3069  pgrad:  0.360757

aspgn <- spg(par = (1:n)/n^2, fn = nllrv, project = proj)

## iter:  0  f-value:  11.3069  pgrad:  0.133333

cat("F_optimal: with gradient=", aspg$value, "  num. approx.=",
    aspgn$value, "\n")
```

```
## F_optimal: with gradient= 8.04719    num. approx.= 8.04719

pbest <- rep(1/n, n)
cat("fbest = ", nllrv(pbest), "  when all parameters = ", pbest[1],
    "\n")

## fbest =  8.04719    when all parameters =  0.2

cat("deviations:  with gradient=", max(abs(aspg$par - pbest)),
    "   num. approx.=", max(abs(aspg$par - pbest)), "\n")

## deviations:  with gradient= 3.81244e-06    num. approx.= 3.81244e-06
```

Here the projection `proj` is the key to success of method `spg`. Other methods do not have the flexibility to impose the projection directly. We would need to carefully build the projection into the function(s) and/or the method codes. This was done by Geradin (1971) for the Rayleigh quotient problem but requires a number of changes to the program code.

13.3   Inequality constraints

Kovalchik et al. (2013) present an interesting and readable approach to estimating what they call a LEXPIT model. This attempts to explain the variation of a binary response variable Y, for example, one that has a 0 for a patient who does not have a particular diseases and 1 when they do. The model uses additive and multiplicative explanatory variables, which are put in matrices X and Z. For explanatory purposes, I will simplify to a single additive covariate x and single multiplicative covariate z. The true but unknown Y is assumed to be distributed as a Bernoulli (i.e., zero–one) variate with probability parameter (a function of x and z)

$$\pi(x, z) = \beta * x + \text{expit}(\text{gamma}_0 + \text{gamma}_1 * z) \tag{13.4}$$

where this probability is clearly bounded by 0 and 1. In fact, this probability should not actually take on those end-of-range values, or we have no variation.

$$0 < \pi(x, z) < 1 \tag{13.5}$$

where

$$\text{expit}(q) = \exp(q)/(1 + \exp(q)) \tag{13.6}$$

is the inverse logit function. The logit function is

$$\text{logit}(p) = \log(p/(1 - p)) \tag{13.7}$$

To illustrate the model and its estimation, let us suppose that we have patients whose age, sugar consumption, and health status according to some 0/1 indicator are known. We now want to model the probability the health (or disease) indicator is 1 by equation (13.4).

We have (actually we generated) 500 observations on status *y*, age and sugar and now want to fit the model by maximum likelihood. We will minimize the negative log likelihood, which is provided by the following code.

```
expit <- function(q) {
    zz <- exp(q)
    zz/(1 + zz)
}

logit <- function(p) {
    log(p/(1 - p))
}

negll <- function(par, data) {
    yy <- data$y
    ss <- data$sugar
    aa <- data$age
    beta0 <- par[1]
    gam0 <- par[2]
    gam1 <- par[3]
    ppi <- beta0 * ss + expit(gam0 + gam1 * aa)
    ## need to watch out for ppi out of bounds
    if (any(ppi >= 1) || any(ppi <= 0)) {
        val <- 1e+200
    } else {
        val <- -sum(log(1 - ppi)) - sum(yy * logit(ppi))
    }
    val
}
```

Kovalchik et al. (2013) suggest that it is useful to provide starting values derived in some way from the data. Here we will take a very crude approach of setting the $\beta = 0$. This implies that we have the simpler logistic regression model, which we can estimate as follows:

```
tglm <- glm(y ~ age, family = binomial(logit), data = mydata)
pars <- c(0, coef(tglm))
names(pars) <- c("beta0", "gam0", "gam1")
pars

##      beta0       gam0       gam1
## 0.0000000 -4.6583342  0.0504476
```

Now we can proceed to fit the LEXPIT model. We will try all the **optimx** methods.

```
require(optimx)
aoptimx <- optimx(pars, negll, "grnd", method = "all",
 control = list(trace = 0), data = mydata)

## Warning: Replacing NULL gr with 'grnd' approximation
## Warning: bounds can only be used with method L-BFGS-B (or Brent)
## Warning: bounds can only be used with method L-BFGS-B (or Brent)
## Warning: bounds can only be used with method L-BFGS-B (or Brent)
```

```
print(summary(aoptimx, order = value))
```

```
##                   beta0     gam0      gam1       value
## Rcgmin       0.1566348 -5.79545 0.0632420  1.99324e+02
## ucminf       0.1566348 -5.79545 0.0632420  1.99324e+02
## Rvmmin       0.1566348 -5.79545 0.0632420  1.99324e+02
## nlminb       0.1566348 -5.79545 0.0632420  1.99324e+02
## nmkb         0.1566367 -5.79532 0.0632398  1.99324e+02
## nlm          0.1565503 -5.79397 0.0632233  1.99324e+02
## spg          0.1565252 -5.79387 0.0632229  1.99324e+02
## hjkb         0.1563568 -5.79060 0.0631810  1.99324e+02
## BFGS         0.0815319 -4.67534 0.0494269  2.00107e+02
## CG           0.0815319 -4.67534 0.0494269  2.00107e+02
## Nelder-Mead  0.0815319 -4.67534 0.0494269  2.00107e+02
## L-BFGS-B     0.0815319 -4.67534 0.0494269  2.00107e+02
## bobyqa       0.0206010 -4.65833 0.0504476  2.00391e+02
## newuoa              NA       NA        NA  8.98847e+307
##              fevals gevals niter convcode  kkt1 kkt2 xtimes
## Rcgmin         1007    150    NA        1  TRUE TRUE  0.736
## ucminf           18     18    NA        0  TRUE TRUE  0.072
## Rvmmin           71     12    NA        0  TRUE TRUE  0.060
## nlminb           29     14    13        0  TRUE TRUE  0.060
## nmkb            217     NA    NA        0  TRUE TRUE  0.040
## nlm              NA     NA    16        0  TRUE TRUE  0.008
## spg            2342     NA  1304        0  TRUE TRUE  5.660
## hjkb           2132     NA    19        0 FALSE TRUE  0.292
## BFGS             13     13    NA        0 FALSE TRUE  0.044
## CG               13     13    NA        0 FALSE TRUE  0.048
## Nelder-Mead      13     13    NA        0 FALSE TRUE  0.052
## L-BFGS-B         13     13    NA        0 FALSE TRUE  0.048
## bobyqa           21     NA    NA        0 FALSE TRUE  0.004
## newuoa           NA     NA    NA     9999    NA   NA  0.004
```

Here we see most of the methods have found a reasonable solution.

We can also try the constrained nonlinear optimization packages **alabama** and **Rsolnp**. First, we need to define our inequality constraints.

```
pibnd <- function(par, data) {
    # inequality constraints
    yy <- data$y
    nobs <- length(yy)
    ss <- data$sugar
    aa <- data$age
    beta0 <- par[1]
    gam0 <- par[2]
    gam1 <- par[3]
    ppi <- beta0 * ss + expit(gam0 + gam1 * aa)
    hh <- rep(NA, 2)
    # Note probability bounded at both 0 and 1
    hh <- c((ppi - 1e-04), (1 - 1e-04 - ppi))
    hh
}

yy <- mydata$y
nobs <- length(yy)
```

```
lo <- rep(1e-04, nobs)
up <- 1 - lo

require(alabama)
coo <- list(trace = FALSE)
aaug <- auglag(par = pars, fn = negll, hin = pibnd, data = mydata,
    control.outer = coo)
aaug$par
```

```
##      beta0        gam0        gam1
##  0.1501910 -5.6903499  0.0619483
```

```
aaug$value
```

```
## [1] 199.33
```

```
aaug$convergence
```

```
## [1] 0
```

```
aco <- constrOptim.nl(par = pars, fn = negll, hin = pibnd, data = mydata,
    control.outer = coo)
aco$par
```

```
##      beta0        gam0        gam1
##  0.1501612 -5.6901151  0.0619458
```

```
aco$value
```

```
## [1] 199.33
```

```
aco$convergence
```

```
## [1] 0
```

Last in this section, we try **Rsolnp**.

```
require(Rsolnp)
pineq <- function(par, data) {
    # we will compute functions that obey the ineqLB and ineqUB
    # constraints
    yy <- data$y
    nobs <- length(yy)
    ss <- data$sugar
    aa <- data$age
    beta0 <- par[1]
    gam0 <- par[2]
    gam1 <- par[3]
    ppi <- beta0 * ss + expit(gam0 + gam1 * aa)
}

asolnp <- solnp(pars = pars, fun = negll, ineqfun = pineq, ineqLB = lo,
    ineqUB = up, data = mydata)
```

```
##
## Iter: 1 fn: 199.3254  Pars:    0.15280 -5.74331  0.06263
## Iter: 2 fn: 199.3239  Pars:    0.15663 -5.79529  0.06324
```

```
## Iter: 3 fn: 199.3239  Pars:    0.15663 -5.79529  0.06324
## solnp--> Completed in 3 iterations

asolnp$pars

##      beta0      gam0      gam1
##   0.156626 -5.795292  0.063240

negll(asolnp$pars, mydata)

## [1] 199.324
```

While we see that several methods have quite satisfactorily solved this estimation problem, I would caution that an important step is that of providing reasonable starting values for the parameters. Constrained problems seem to require better starts than their unconstrained counterparts, but this is an opinion based on limited experience rather than exhaustive evidence.

13.4   A perspective on penalty function ideas

The logarithmic barrier function is generally used to make the objective function so large that the parameters are deflected away from a constraint. An alternative is to penalize violation of the constraint. This actually allows the parameters to take on values that are infeasible, that is, outside the bounds. Increasing a weight or control on the penalty allows us to make it more stringent, and generally the parameters become "less infeasible" in such approaches, if we can tolerate such linguistic torture.

An annoyance with penalty methods is that we may introduce very large numbers as we try to force feasibility, and this may degrade the performance of our (unconstrained) optimization methods. In my own work, I have found penalty methods more useful as a trial to learn about problems than as an ongoing tool for solving a family of optimization problems. Barrier functions, as we have seen, can be used if we start and finish inside the feasible region.

13.5   Assessment

In this chapter, we have seen that there are several ways that R can be used to tackle nonlinear optimization problems with general constraints. What are the lessons to take away? Here are my views:

- As a first try, it is best to stick with the problem as stated and verify the objective and constraints.

- Keep any scaling simple, such as powers of 10. This can help to render the problem less difficult for the optimization software, and it does not involve complicated algebra which may be error prone.

- Equality constraints can be surprisingly difficult to deal with. I prefer, if possible, to eliminate one or more parameters, but as we have seen, one needs to consider which parameter(s) to eliminate.

- Penalty methods are simple and useful for exploring a problem, especially if infeasible parameters can be tolerated. Otherwise, barrier methods will keep parameters feasible, but we do need an initial feasible set of parameters.

- Good starting parameters are important to success.

- As a continuing check, evaluate the objective and constraints in the original specification of the problem.

References

Geradin M 1971 The computational efficiency of a new minimization algorithm for eigenvalue analysis. Journal of Sound and Vibration 19, 319–331.

Kovalchik SA, Varadhan R, Fetterman B, Poitras NE, Wacholder S, and Katki HA 2013 A general binomial regression model to estimate standardized risk differences from binary response data. Statistics in Medicine 32(5), 808–821.

Nash JC 1979 Compact Numerical Methods for Computers: Linear Algebra and Function Minimisation. Adam Hilger, Bristol. Second Edition, 1990, Bristol: Institute of Physics Publications.

14

Applications of mathematical programming

Mathematical programming is, in my view, satisfaction of constraints with an attempt to minimize or maximize an objective function. That is, there are "many" constraints and they are the first focus of our attention. Moreover, the subject has been driven by a number of practical-resource-assignment problems, so the variables may also be integers instead of or as well as real numbers.

14.1 Statistical applications of math programming

In this field, R shows its statistical heritage and is not at the forefront. Most statisticians do not use math programming. I include myself in this. My needs for such tools have largely been driven by approximation problems where we wish to minimize the sum of absolute errors or the maximum absolute error. There are also situations where I have wanted to solve some optimization problems such as the diet problem, where the constraints are the required total intake of nutrients or limits on salt and fats and the objective is to minimize the cost to provide the diet.

In the 1980s, the University of New South Wales ran a program directed by Bruce Murtagh to offer companies student help with optimization problems at bargain consulting prices. I believe that $5000 was paid by a cat food company to determine an economical blend for tinned pet food. The resulting mixture, determined from a linear program, used ground up fish heads and fish oil with mostly grain components for a saving many times the consulting fee. This is not the sort of problem R users generally solve.

There are, however, some applications to statistics. One area is that of obfuscating published statistics to render difficult or impossible the retrieval of private information by subtraction from totals. See, for example, the Statistics Canada report "Data Quality and Confidentiality Standards and Guidelines (Public): 2011 Census Dissemination" (http://www12.statcan.gc.ca/census-recensement/2011/ref/DQ-QD/2011_DQ-QD_Guide_E.pdf). Gordon Sande presents some ideas on this in a paper to the 2003 conference of the (US) Federal Committee on Statistical Methodology (http://www.fcsm.gov/events/papers2003.html). The idea is to use constraints to keep published data far enough from actual figures to inhibit the revelation of individual data by calculation, while minimizing the perturbations in some way. Another use of math programming in large statistical agencies is presented by McCarthy and Bateman (1988) who consider the design of surveys. Bhat (2011) presents a quite good overview of math programming in statistics in his doctoral thesis introduction, listing some applications in contingency tables and hypothesis testing as well as the minimax and least absolute deviations problems that I mentioned earlier. There are also some applications in machine learning, for example, http://stackoverflow.com/questions/12349122/solving-quadratic-programming-using-r.

14.2   R packages for math programming

We have already dealt with the general nonlinear programming problem in Chapter 13. However, my experience has been that despite the formal statement of the problem – minimize a nonlinear objective subject to possibly nonlinear constraints – is the generalization of traditional linear and quadratic programming problems of this chapter, many problems have rather few constraints. They are constrained nonlinear function minimization problems.

In Chapter 13, we considered both **alabama** (Varadhan, 2012) and **Rsolnp** (Ghalanos and Theussl, 2012). These packages require, in my view, more care and attention to use than most discussed in this book. **Rsolnp** is the only package that is listed in the R-forge project "Integer and Nonlinear Optimization in R (RINO)", but there are three other packages in the code repository. However, none of these has been updated in 3 years before the time of writing, while **Rsolnp** appears to be maintained. Because the packages in this paragraph have more kinship with function minimization, we will not address them further in this chapter.

Most traditional math programming involves some special form of objective or constraints, in particular linear constraints with either a linear or a quadratic objective. This leads to the well-established **linear programming** (LP) and **quadratic programming** problems.

Quadratic programming has two tools as the main candidates in R. Solve.QP() from package **quadprog** (original by Berwin A. Turlach R port by Andreas Weingessel, 2013) has to my eyes the more traditional interface, while ipop() from package **kernlab** (Karatzoglou et al. 2004) requires a somewhat unusual format for specifying the linear constraints.

For linear programming, packages **linprog** (Henningsen, 2012) and **lpSolve** (Berkelaar et al. 2013) have slightly different calling sequences. Both of these are for real variables. However, much of LP is for integer or mixed-integer variables. The package **Rglpk** (Theussl and Hornik, 2013) provides an interface to the Gnu LP kit and can handle both real- and mixed-integer problems. The same authors provide **Rsymphony** (Harter et al. 2013) and **ROI** (Hornik et al. 2011) for mixed-integer programming problems. These are all quite complicated packages, with a style for inputting a problem that is quite different from the problem statements we have seen elsewhere in this book.

Similarly, Jelmer Ypma of the University College London offers interfaces **nloptr** to Steven Johnson's NLopt and **ipoptr** to the COIN-OR interior point optimizer(s) `ipopt()`. (This should not be confused with `ipop()` from **kernlab**.) These are not part of the CRAN repository and like the tools of the previous paragraph require installation in a way that is not common with other R packages. I have decided not to discuss either set of tools as I have not had enough experience with them to comment fairly. However, it is important to note that interior point methods have been much discussed in the optimization literature over the past two decades, and there is no mainstream R package that addresses the need. Nor does R appears to offer any tool for convex optimization at the time of writing.

There are also R interfaces to some commercial solvers. **Rcplex** is an interface to the CPLEX optimization engine. The history of this suite of solvers is described in http://en.wikipedia.org/wiki/CPLEX. The R interface appears to support only Linux platforms. Similarly, **Rmosek** (http://rmosek.r-forge.r-project.org/) is an R interface to the MOSEK solvers and claims to deal with large-scale optimization problems having linear, quadratic, and some other objectives with integer constraints. Again, installation will be nonstandard because the user must first install the commercial package. I have no experience with such packages and will not provide a treatment here.

There are R packages that build on some of the tools above. For example, package **parma** for "Portfolio Allocation and Risk Management Applications" calls the **Rglpk** tools. In the rest of this chapter, we will look at using some of them for some illustrative problems of a statistical nature.

14.3   Example problem: L1 regression

Traditional regression estimation problems minimize the sum of squared residuals. However, there are many arguments in favor of other loss functions based on the residuals (L-infinity or minimax regression, M-estimators, and others). Here we will look at L1 or least absolute deviations (LAD) regression. For a linear model, this is stated as

$$x^* = \mathrm{argmin}_x(\mathrm{sum}(\mathrm{abs}(Ax - y))) \qquad (14.1)$$

where A is the matrix of predictor variables (generally one column is all 1's to provide a constant or intercept term) and y is the independent or predicted variable. Instead of minimizing the sum of squared residuals, we are minimizing the sum of absolute values.

Let us set up a small problem.

```
x <- c(3, -2, -1)
a1 <- rep(1, 10)
a2 <- 1:10
a3 <- a2^2
A <- matrix(c(a1, a2, a3), nrow = 10, ncol = 3)
A
```

```
##         [,1] [,2] [,3]
##  [1,]    1    1    1
##  [2,]    1    2    4
##  [3,]    1    3    9
##  [4,]    1    4   16
##  [5,]    1    5   25
##  [6,]    1    6   36
##  [7,]    1    7   49
##  [8,]    1    8   64
##  [9,]    1    9   81
## [10,]    1   10  100
```

```
## create model data
y0 = A %*% x
y0
```

```
##         [,1]
##  [1,]     0
##  [2,]    -5
##  [3,]   -12
##  [4,]   -21
##  [5,]   -32
##  [6,]   -45
##  [7,]   -60
##  [8,]   -77
##  [9,]   -96
## [10,]  -117
```

```
## perturb the model data --- get working data
set.seed(1245)
y <- as.numeric(y0) + runif(10, -1, 1)
y
```

```
## [1]    0.324886   -4.428305  -12.854561  -20.393083
## [5]  -32.854574  -45.193280  -60.690202  -77.243728
## [9]  -96.406035 -117.241829
```

```
## try a least squares model
l2fit <- lm(y ~ a2 + a3)
l2fit
```

```
##
## Call:
## lm(formula = y ~ a2 + a3)
##
## Coefficients:
## (Intercept)           a2           a3
##       3.737       -2.339       -0.976
```

It turns out that our regular optimizers get into trouble solving this in the L1 norm. We also check the L2 or square norm solution, which turns out to be much less trouble.

```
sumabs <- function(x, A, y) {
    # loss function
    sum(abs(y - A %*% x))
}
sumsqr <- function(x, A, y) {
    # loss function
    sum((y - A %*% x)^2) # a poor way to do this
}
require(optimx)
st <- rep(1, 3)
ansL1 <- optimx(st, sumabs, method = "all", control = list
(usenumDeriv = TRUE), A = A, y = y)

## Warning: Replacing NULL gr with 'numDeriv' approximation
## Warning: bounds can only be used with method L-BFGS-B (or Brent)
## Warning: bounds can only be used with method L-BFGS-B (or Brent)
## Warning: bounds can only be used with method L-BFGS-B (or Brent)

print(summary(ansL1, order = value))

##                    p1         p2         p3        value
## nmkb         3.605201  -2.302050  -0.978265   3.29511e+00
## nlm          3.582684  -2.277278  -0.980520   3.29524e+00
## ucminf       3.545880  -2.236248  -0.984252   3.29539e+00
## hjkb         3.490341  -2.175003  -0.989822   3.29551e+00
## Rvmmin       3.604006  -2.290916  -0.979367   3.30003e+00
## nlminb       3.509092  -2.210676  -0.986442   3.31376e+00
## bobyqa       3.418332  -2.049762  -1.004648   3.77613e+00
## BFGS         0.352885  -1.263250  -1.054194   7.07201e+00
## CG           0.352885  -1.263250  -1.054194   7.07201e+00
## Nelder-Mead  0.352885  -1.263250  -1.054194   7.07201e+00
## L-BFGS-B     0.352885  -1.263250  -1.054194   7.07201e+00
## newuoa       0.427852  -0.821233  -1.104232   9.20869e+00
## Rcgmin       0.902925   0.542689  -1.288875   2.50204e+01
## spg                NA         NA         NA   8.98847e+307
##          fevals gevals niter convcode   kkt1   kkt2 xtimes
## nmkb        257     NA    NA        0  FALSE   TRUE  0.032
## nlm          NA     NA    29        0  FALSE  FALSE  0.004
## ucminf       51     51    NA        0  FALSE  FALSE  0.008
## hjkb       6764     NA    19        0  FALSE   TRUE  0.108
## Rvmmin      142     20    NA        0  FALSE  FALSE  0.012
## nlminb       54     22    22        1  FALSE  FALSE  0.004
## bobyqa      206     NA    NA        0  FALSE  FALSE  0.004
## BFGS        146    146    NA        0  FALSE  FALSE  0.016
## CG          146    146    NA        0  FALSE  FALSE  0.012
## Nelder-Mead 146    146    NA        0  FALSE  FALSE  0.016
## L-BFGS-B    146    146    NA        0  FALSE  FALSE  0.016
## newuoa      302     NA    NA        0  FALSE   TRUE  0.004
## Rcgmin     1011    112    NA        1  FALSE  FALSE  0.036
## spg          NA     NA    NA     9999     NA     NA  0.160

ansL2 <- optimx(st, sumsqr, method = "all", control = list
(usenumDeriv = TRUE), A = A, y = y)
```

```
## Warning: Replacing NULL gr with 'numDeriv' approximation
## Warning: bounds can only be used with method L-BFGS-B (or Brent)
## Warning: bounds can only be used with method L-BFGS-B (or Brent)
## Warning: bounds can only be used with method L-BFGS-B (or Brent)

print(summary(ansL2, order = value))
```

```
##                  p1       p2        p3        value fevals
## Rvmmin      3.73671 -2.33884 -0.975874  1.91298e+00     92
## nlm         3.73671 -2.33884 -0.975874  1.91298e+00     NA
## ucminf      3.73671 -2.33884 -0.975874  1.91298e+00     10
## nlminb      3.73671 -2.33884 -0.975874  1.91298e+00      9
## BFGS        3.73671 -2.33884 -0.975874  1.91298e+00     18
## CG          3.73671 -2.33884 -0.975874  1.91298e+00     18
## Nelder-Mead 3.73671 -2.33884 -0.975874  1.91298e+00     18
## L-BFGS-B    3.73671 -2.33884 -0.975874  1.91298e+00     18
## Rcgmin      3.73671 -2.33884 -0.975874  1.91298e+00   1007
## newuoa      3.73671 -2.33884 -0.975874  1.91298e+00    273
## bobyqa      3.73670 -2.33884 -0.975874  1.91298e+00    612
## hjkb        3.73667 -2.33882 -0.975876  1.91298e+00   2600
## nmkb        3.73648 -2.33865 -0.975894  1.91298e+00    150
## spg              NA       NA        NA  8.98847e+307    NA
##               gevals niter convcode  kkt1 kkt2 xtimes
## Rvmmin            19    NA        0  TRUE TRUE  0.000
## nlm               NA    10        0  TRUE TRUE  0.000
## ucminf            10    NA        0  TRUE TRUE  0.000
## nlminb             8     7        0  TRUE TRUE  0.000
## BFGS              18    NA        0  TRUE TRUE  0.000
## CG                18    NA        0  TRUE TRUE  0.004
## Nelder-Mead       18    NA        0  TRUE TRUE  0.004
## L-BFGS-B          18    NA        0  TRUE TRUE  0.000
## Rcgmin           149    NA        1  TRUE TRUE  0.040
## newuoa            NA    NA        0  TRUE TRUE  0.004
## bobyqa            NA    NA        0  TRUE TRUE  0.008
## hjkb              NA    19        0  TRUE TRUE  0.040
## nmkb              NA    NA        0 FALSE TRUE  0.016
## spg               NA    NA     9999    NA   NA  0.164
```

```
## l2fit sum of squares
sumsqr(coef(l2fit), A, y)
```

```
## [1] 1.91298
```

We can, however, use LP to solve the LAD problem in m observations and n parameters as defined by the A matrix and y vector above. Following the treatment in Bisschop and Roelofs (2006), we establish the problem as a minimization of

$$\sum_{i=1}^{m} z_i \qquad (14.2)$$

subject to the constraints

$$z \geq y - Ax \qquad (14.3)$$

$$z \geq -y + Ax \qquad (14.4)$$

These strange-looking constraints, imposed simultaneously, push the sum of absolute residuals to the minimum.

Unfortunately, most LP solvers automatically impose nonnegativity constraints on the x variables that we wish to find. We work around this by defining

$$x = u - v \tag{14.5}$$

where the u and v variables are nonnegative. Furthermore, the solver we will call here, SolveLP() from package **linprog**, prefers the constraints to be as $\leq$ rather than $\geq$ forms. To make this work, we build a supermatrix

$$A_{\text{big}} = \begin{pmatrix} -I & -A & A \\ -I & A & -A \end{pmatrix} \tag{14.6}$$

The right-hand side of the constraints is then the vector (bvec)

$$\begin{pmatrix} -y \\ y \end{pmatrix} \tag{14.7}$$

and the objective function is defined by a vector (cvec) that has m ones and $2 * n$ zeros. The latter elements are used to define the u and v elements of our solution, while the ones are to give us the objective function.

This setup is captured in the following function, called LADreg, which takes as input the matrix A and the vector y.

```
LADreg <- function(A, y) {
    # Least sum abs residuals regression
    require(linprog)
    m <- dim(A)[1]
    n <- dim(A)[2]
    if (length(y) != m)
        stop("Incompatible A and y sizes in LADreg")
    ones <- diag(rep(1, m))
    Abig <- rbind(cbind(-ones, -A, A), cbind(-ones, A, -A))
    bvec <- c(-y, y)
    cvec <- c(rep(1, m), rep(0, 2 * n)) # m z's, n pos x, n neg x
    ans <- solveLP(cvec, bvec, Abig, lpSolve = TRUE)
    sol <- ans$solution
    coeffs <- sol[(m + 1):(m + n)] - sol[(m + n + 1):(m + 2 * n)]
    names(coeffs) <- as.character(1:n)
    res <- as.numeric(sol[1:m])
    sad <- sum(abs(res))
    answer <- list(coeffs = coeffs, res = res, sad = sad)
}
```

Let us use this to solve our example problem.

```
myans1 <- LADreg(A, y)
myans1$coeffs

##          1         2         3
##  3.489517 -2.174797 -0.989834

myans1$sad

## [1] 3.2951
```

It turns out that there are tools in R that are more specifically intended for this problem. In particular, L1linreg() from package **pracma** and rq() from **quantreg** can do the job. **quantreg** (Koenker, 2013) also has a nonlinear quantile regression function nlrq(). Koenker (2005) has a large presence in the area of quantile regression and his package is well established and can be recommended for this application. The *ad hoc* treatment above is intended to provide an introduction to how similar problems can be approached with available tools.

```
require(pracma)
ap1 <- L1linreg(A = A, b = y)
ap1

## $x
## [1]  3.605010 -2.301839 -0.978285
##
## $reltol
## [1] 2.41727e-08
##
## $niter
## [1] 20

sumabs(ap1$x, A, y)

## [1] 3.2951

require(quantreg)
qreg <- rq(y ~ a2 + a3, tau = 0.5)

## Warning: Solution may be nonunique

qreg

## Call:
## rq(formula = y ~ a2 + a3, tau = 0.5)
##
## Coefficients:
## (Intercept)          a2          a3
##    3.489517   -2.174797   -0.989834
##
## Degrees of freedom: 10 total; 7 residual

xq <- coef(qreg)
sumabs(xq, A, y)

## [1] 3.2951
```

14.4   Example problem: minimax regression

A very similar problem is the minimax or L-infinity regression problem. For a linear model, this is stated

$$x^* = \text{argmin}_x(\max(\text{abs}(Ax - y))) \tag{14.8}$$

where A is the matrix of predictor variables (generally, one column is all 1's to provide a constant or intercept term) and y is the independent or predicted variable. That is, we replace the sum with maximum in finding the measure of the size of the absolute residuals.

Once again, our traditional optimization tools do poorly. We will use the data as above.

```
minmaxw <- function(x, A, y) {
    res <- y - A %*% x
    val <- max(abs(res))
    attr(val, "where") <- which(abs(res) == val)
    val
}
minmax <- function(x, A, y) {
    res <- y - A %*% x
    val <- max(abs(res))
}
require(optimx)
```

```
## Loading required package: optimx
```

```
st <- c(1, 1, 1)
ansmm1 <- optimx(st, minmax, gr = "grnd", method = "all", A = A,
    y = y)
```

```
## Warning: Replacing NULL gr with 'grnd' approximation
## Loading required package: numDeriv
## Warning: bounds can only be used with method L-BFGS-B (or Brent)
## Warning: bounds can only be used with method L-BFGS-B (or Brent)
## Warning: bounds can only be used with method L-BFGS-B (or Brent)
```

```
print(summary(ansmm1, order = value))
```

```
##                   p1          p2          p3      value fevals
## ucminf       2.783021  -1.9519752  -1.00600  0.727790     49
## Rvmmin       2.296971  -1.7319634  -1.02950  0.856668     73
## nmkb         2.361915  -1.8630677  -1.01989  1.015456    375
## spg          0.893647  -0.9462155  -1.10152  1.479067   5550
## nlm          0.914282  -0.9066519  -1.10660  1.570571     NA
## hjkb         0.424068  -0.6879692  -1.12500  1.713797    446
## nlminb       0.503606  -1.2808044  -1.02808  2.130163     77
## bobyqa       3.194253  -0.0163896  -1.25020  4.747844     82
## BFGS         0.974467   0.7610416  -1.30768  4.942183     96
## CG           0.974467   0.7610416  -1.30768  4.942183     96
## Nelder-Mead  0.974467   0.7610416  -1.30768  4.942183     96
## L-BFGS-B     0.974467   0.7610416  -1.30768  4.942183     96
## Rcgmin       0.975069   0.7629142  -1.30792  4.946218   1013
## newuoa       2.817482   0.5573549  -1.31273  5.640493     79
##                gevals niter convcode  kkt1  kkt2 xtimes
```

```
## ucminf        49    NA      0 FALSE  TRUE  0.040
## Rvmmin        15    NA     -1 FALSE FALSE  0.012
## nmkb          NA    NA      0 FALSE FALSE  0.024
## spg           NA  1413      0 FALSE FALSE  1.284
## nlm           NA    29      0 FALSE FALSE  0.000
## hjkb          NA    19      0 FALSE  TRUE  0.008
## nlminb        28    28      1 FALSE FALSE  0.020
## bobyqa        NA    NA      0 FALSE  TRUE  0.000
## BFGS          96    NA     52 FALSE FALSE  0.072
## CG            96    NA     52 FALSE FALSE  0.068
## Nelder-Mead   96    NA     52 FALSE FALSE  0.068
## L-BFGS-B      96    NA     52 FALSE FALSE  0.064
## Rcgmin        97    NA      1 FALSE FALSE  0.092
## newuoa        NA    NA      0 FALSE  TRUE  0.000
```

The setup for solving the problem by LP is rather similar to that for LAD regression (Bisschop and Roelofs, 2006), but this time there is but the single largest residual to minimize. The trick is that we need to choose which one of the set of residuals will be the largest in absolute value. To do this, we constrain the residuals to be smaller than this element across all residuals. The following code computes the solution by a traditional linear program.

```
mmreg <- function(A, y) {
    # max abs residuals regression
    require(linprog)
    m <- dim(A)[1]
    n <- dim(A)[2]
    if (length(y) != m)
        stop("Incompatible A and y sizes in LADreg")
    onem <- rep(1, m)
    Abig <- rbind(cbind(-A, A, -onem), cbind(A, -A, -onem))
    bvec <- c(-y, y)
    cvec <- c(rep(0, 2 * n), 1) # n pos x, n neg x, 1 t
    ans <- solveLP(cvec, bvec, Abig, lpSolve = TRUE)
    sol <- ans$solution
    coeffs <- sol[1:n] - sol[(n + 1):(2 * n)]
    names(coeffs) <- as.character(1:n)
    res <- as.numeric(A %*% coeffs - y)
    mad <- max(abs(res))
    answer <- list(coeffs = coeffs, res = res, mad = mad)
}
```

Let us apply this to our problem.

```
amm <- mmreg(A, y)
amm

## $coeffs
##        1         2         3
##  2.73966  -1.92954  -1.00881
##
## $res
##  [1] -0.5235670 -0.7263380  0.7263380 -0.7263380  0.7263380
##  [6]  0.0386108  0.4914828 -0.0166592  0.0663623 -0.1947466
```

```
##
## $mad
## [1] 0.726338

mmcoef <- amm$coeffs
## sum of squares of minimax solution
print(sumsqr(mmcoef, A, y))

## [1] 2.67004

## sum abs residuals of minimax solution
print(sumabs(mmcoef, A, y))

## [1] 4.23678
```

14.5   Nonlinear quantile regression

Koenker's **quantreg** has function `nlrq()` that parallels the functionality of the nonlinear least squares tools `nls()`, `nlxb()`, and `nlsLM()`. Let us apply this to our scaled Hobbs problem (Section 1.2).

```
y <- c(5.308, 7.24, 9.638, 12.866, 17.069, 23.192, 31.443, 38.558,
    50.156, 62.948, 75.995, 91.972)
t <- 1:12
weeddata <- data.frame(y = y, t = t)
require(quantreg)
formulas <- y ~ 100 * sb1/(1 + 10 * sb2 * exp(-0.1 * sb3 * t))
snames <- c("sb1", "sb2", "sb3")
st <- c(1, 1, 1)
names(st) <- snames
anlrqsh1 <- nlrq(formulas, start = st, data = weeddata)
summary(anlrqsh1)

##
## Call: nlrq(formula = formulas, data = weeddata, start = st,
control = structure(list(
##    maxiter = 100, k = 2, InitialStepSize = 1, big = 1e+20, eps = 1e-07,
##    beta = 0.97), .Names = c("maxiter", "k", "InitialStepSize",
## "big", "eps", "beta")), trace = FALSE)
##
## tau: [1] 0.5
##
## Coefficients:
##        Value      Std. Error   t value     Pr(>|t|)
## sb1 86710.77088 10616.02767    8.16791     0.00002
## sb2 98665.17861     0.00000        Inf     0.00000
## sb3     1.95666     0.14804   13.21723     0.00000
```

The coefficients here, found with surprisingly few iterations, are clearly very unlike the answers found in Section 16.7. We try again from a more sensible set of starting parameters and get a result quite close to the least squares answer.

```
st <- c(2, 5, 3)
names(st) <- snames
anlrqsh2 <- nlrq(formulas, start = st, data = weeddata)
summary(anlrqsh2)

##
## Call: nlrq(formula = formulas, data = weeddata, start = st,
control = structure(list(
##      maxiter = 100, k = 2, InitialStepSize = 1, big = 1e+20, eps = 1e-07,
##      beta = 0.97), .Names = c("maxiter", "k", "InitialStepSize",
## "big", "eps", "beta")), trace = FALSE)
##
## tau: [1] 0.5
##
## Coefficients:
##      Value    Std. Error  t value   Pr(>|t|)
## sb1  1.94496  0.22886     8.49839   0.00001
## sb2  4.95160  0.48881    10.12981   0.00000
## sb3  3.16140  0.08643    36.57837   0.00000
```

As with most nonlinear modeling, L1 regression or quantile regression where we set the quantile parameter tau = 0.5 requires that we pay attention to starting values and the possibility of multiple solutions. That said, I find the package **quantreg** to be very well done.

14.6   Polynomial approximation

A particular problem that arises from time to time is to develop a polynomial approximation to a function that has known maximum error (see, for example, http://en.wikipedia.org/wiki/Minimax_approximation_algorithm). Note that there is a well-developed literature on this subject, and the Remez algorithm is the gold standard for this task (see Pachón and Trefethen (2009) for a recent contribution to this topic). Unfortunately, R appears to lack effective methods for this approach.

If our function is $f(x)$ and we want a polynomial in degree $(n - 1)$, this gives the problem of finding the coefficients of the vector b in

$$b^* = \text{argmin}_b(\text{abs}(f(x) - \sum_{j=1}^{n} b_j x^{j-1})) \tag{14.9}$$

for all x in some interval $l <= x <= u$.

Note that the goal is to approximate a **function** with a polynomial of given degree, and this is where the Remez approach is the one to use.

There is a discretized version of this problem where we attempt to find the solution over a discrete set of values of x. Here is an example problem suggested by Hans Werner Borchers (Figure 14.1). Let us first get the least squares fit, which will provide some idea of where the solution may lie.

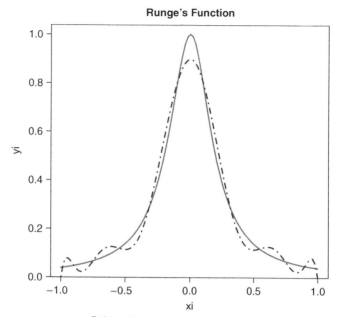

Figure 14.1

```
n <- 10 # polynomial of degree 10
m <- 101 # no. of data points
xi <- seq(-1, 1, length = m)
yi <- 1/(1 + (5 * xi)^2)
plot(xi, yi, type = "l", lwd = 3, col = "gray", main = "Runge's function")
title(sub = "Polynomial approximation is dot-dash line")

pfn <- function(p) max(abs(polyval(c(p), xi) - yi))

require(pracma)
pf <- polyfit(xi, yi, n)
print(pfn(pf))

## [1] 0.102366

lines(xi, polyval(pf, xi), col = "blue", lty = 4)
```

From the graph, the solution is clearly not a terribly good approximation, but may nonetheless serve us if we know the maximum absolute error over the 101 points in the interval $[-1, 1]$.

```
## set up a matrix for the polynomial data
ndeg <- 10
Ar <- matrix(NA, nrow = 101, ncol = ndeg + 1)
Ar[, ndeg + 1] <- 1
for (j in 1:ndeg) {
```

```
    # align to match polyval
    Ar[, ndeg + 1 - j] <- xi * Ar[, ndeg + 2 - j]
}
cat("Check minimax function value for least squares fit\n")

## Check minimax function value for least squares fit

print(minmaxw(pf, Ar, yi))

## [1] 0.102366
## attr(,"where")
## [1] 51

apoly1 <- mmreg(Ar, yi)
apoly1$mad

## [1] 0.0654678

pcoef <- apoly1$coeffs
pcoef

##            1          2          3          4          5
##   -50.248256   0.000000 135.853517   0.000000 -134.201071
##            6          7          8          9         10
##     0.000000  59.193155   0.000000 -11.558883   0.000000
##           11
##     0.934532

sumabs(pcoef, Ar, yi)

## [1] 4.28865

sumsqr(pcoef, Ar, yi)

## [1] 0.224028

print(minmaxw(pcoef, Ar, yi))

## [1] 0.0654678
## attr(,"where")
## [1]  3 99


## LAD regression on Runge
apoly2 <- LADreg(Ar, yi)
apoly2$sad

## [1] 3.06181

pcoef2 <- apoly2$coeffs
pcoef2

##              1            2            3            4
## -2.19324e+01  0.00000e+00  6.29108e+01  0.00000e+00
##              5            6            7            8
## -6.78485e+01  0.00000e+00  3.41440e+01 -1.39697e-12
##              9           10           11
## -8.11899e+00  4.66635e-14  8.45327e-01
```

```
sumabs(pcoef2, Ar, yi)

## [1] 3.06181

sumsqr(pcoef2, Ar, yi)

## [1] 0.211185

print(minmaxw(pcoef2, Ar, yi))

## [1] 0.154673
## attr(,"where")
## [1] 51
```

We can check the solution if we have a usable global minimization tool. It turns out that the usual stochastic tools (see chapter 15) do rather poorly on this problem, but package **adagio** does find the same solution, albeit slowly.

```
require(adagio)
system.time(aadag <- pureCMAES(pf, pfn, rep(-200, 11), rep(200,
    11)))

##     user   system  elapsed
##   32.110    1.684   33.815

aadag

## $xmin
##  [1] -5.02483e+01 -7.66372e-14  1.35854e+02  1.49894e-14
##  [5] -1.34201e+02 -5.40655e-14  5.91932e+01  1.07589e-13
##  [9] -1.15589e+01 -1.04359e-14  9.34532e-01
##
## $fmin
## [1] 0.0654678
```

References

Berkelaar M et al. 2013 *lpSolve: interface to Lp_solve v. 5.5 to solve linear / integer programs*. R package version 5.6.7.

Bhat KA 2011 Some aspects of mathematical programming in statistics PhD thesis Post Graduate, PhD Dissertation, Department of Statistics, University of Kashmir Srinagar, India.

Bisschop J and Roelofs M 2006 AIMMS Optimization Modeling. Paragon Decision Technology in Haarlem, The Netherlands.

Ghalanos A and Theussl S 2012 *Rsolnp: General Non-linear Optimization Using Augmented Lagrange Multiplier Method*. R package version 1.14.

Harter R, Hornik K and Theussl S 2013 *Rsymphony: Symphony in R*. R package version 0.1-16.

Henningsen A 2012 *linprog: Linear Programming / Optimization*. R package version 0.9-2.

Hornik K, Meyer D and Theussl S 2011 *ROI: R Optimization Infrastructure*. R package version 0.0-7.

Karatzoglou A, Smola A, Hornik K, and Zeileis A 2004 kernlab – an S4 package for kernel methods in R. Journal of Statistical Software 11(9), 1–20.

Koenker R 2005 *Quantile Regression*. Cambridge University Press, Cambridge.

Koenker R 2013 *quantreg: Quantile Regression*. R package version 5.05.

McCarthy WF and Bateman DV 1988 The use of mathematical programming for designing dual frame surveys. *Proceedings of the Survey Research Methods Section, 1988*, pp. 652–653. American Statistical Association, Alexandria, VA.

Pachón R and Trefethen LN 2009 Barycentric-Remez algorithms for best polynomial approximation in the chebfun system. BIT Numerical Mathematics 49(4), 721–741.

Theussl S and Hornik K 2013 *Rglpk: R/GNU Linear Programming Kit Interface*. R package version 0.5-1.

Turlach BA R port by Andreas Weingessel S 2013 *quadprog: Functions to solve Quadratic Programming Problems*. R package version 1.5-5.

Varadhan R 2012 *alabama: Constrained nonlinear optimization*. R package version 2011.9-1.

15

Global optimization and stochastic methods

This chapter looks at some tools designed for those awkward problems with potentially many local extrema – we will again present the ideas by talking of minima. Generally, but not exclusively, R approaches these types of problems stochastically. That is, we try many starting points and find many local minima, then choose the "best."

15.1 Panorama of methods

For the person simply wishing to find a solution to a problem, the fields of global and stochastic optimization are, sadly, overpopulated with methods. Worse, although I have neither had the time nor the inclination to thoroughly research the subject, it appears that there is much rehashing of ideas, so that many of the "new method" papers and programs are really not new. Indeed, one could suggest that the authors of many papers have more in common with the marketers of laundry detergent than mathematicians.

The names associated with these stochastic methods sometimes reflect the philosophical motivation: Simulated Annealing, Genetic Algorithms, Tabu Search, differential evolution, particle swarm optimization, Self-Organizing Migrating Algorithm. The fact that there are so many approaches and methods speaks loudly of the difficulty of such problems. Moreover, it is exceedingly difficult to compare such methods without conducting large-scale tests from different starting points. Guidance to users is therefore limited to the general comment that stochastic methods can have widely varying success depending on the particular problems presented.

Nonlinear Parameter Optimization Using R Tools, First Edition. John C. Nash.
© 2014 John Wiley & Sons, Ltd. Published 2014 by John Wiley & Sons, Ltd.
Companion Website: www.wiley.com/go/nonlinear_parameter

15.2   R packages for global and stochastic optimization

R has many packages and tools in this area. I cannot claim to have tried more than a small sample – I just do not have the time to do so, and packages each have their own setup and calling sequence. Here, however, is a brief overview. I have tried to organize the list so similar packages are together.

- *SANN()* in optim(): A simulated annealing method that uses ONLY the iteration count as a termination criterion. I do not recommend this tool for that reason.

- *GenSA*: Generalized simulated annealing. Unlike the optim() method "SANN," the function GenSA() includes several controls for terminating the optimization process (Xiang et al. 2013).

- *likelihood* uses simulated annealing as its optimization tool. Unfortunately, at the time of writing, this is particularized to the objective functions of the package, but the author (Lora Murphy) has shown interest in generalizing the optimizer, which is written in R (Murphy, 2012).

- *GrassmannOptim*: This package for Grassmann manifold optimization (used in image recognition) uses simulated annealing for attempting global optimization. The documentation (I have not tried this package and have mentioned it to show the range of possibilities) suggests the package can be used for general optimization (Adragni et al. 2012).

- *DEoptim*: Global optimization by differential evolution. This package has many references, but here we will choose to point to Mullen et al. (2011).

- *RcppDE*: Global optimization by differential evolution in C++, aimed at a more efficient interface to the DEoptim functionality (Eddelbuettel, 2012).

- *rgenoud*: R version of genetic optimization uUsing derivatives (Mebane and Sekhon, 2011).

- *GA*: Genetic algorithms – a set of tools for maximizing a fitness function (Scrucca, 2013). A number of examples are included, including minimization of a two-dimensional Rastrigin function (we use a five-dimensional example in the following).

- *gaoptim*: Another suite of genetic algorithms. This one has functions to set up the process and create a function that is used to evolve the solution. However, I did not find that this package worked at all well for me, although I suspect I may have misunderstood the documentation.

- *Rmalschains*: continuous optimization using memetic algorithms with local search chains (MA-LS-Chains) in R (Bergmeir et al. 2012). I confess to knowing rather little about this approach.

- *smco*: A simple Monte Carlo optimizer using adaptive coordinate sampling (de Colombia, 2012).

- *soma*: General-purpose optimization with the self-organizing migrating algorithm (based on the work of Ivan Zelinka, 2011). Again, this approach is one with which I am not familiar.

There are also several other packages that might be used for stochastic optimization rgp (an R genetic programming framework) likely has many tools that could be used for stochastic optimization, but the structure of the package makes it difficult to modify existing scripts to the task. The documentation of package (Elitist nondominated sorting genetic algorithm based on R) was, to my view, such that I could not figure out how to use it to minimize a function. Package **mcga** (machine-coded genetic algorithms for real-valued optimization problems) appears to recode the parameters to 8-bit values and finds the minimum of the encoded function. However, it took me some time to decide that the "optimization" was a minimization. Many genetic and related codes seek maxima.

15.3   An example problem

To illustrate some of these packages, we will use a well-known multimodal test function called the Rastrigin function (Mühlenbein et al. 1991; Törn and Zilinskas, 1989). This is defined in the examples for package **GenSA** as follows.

```
Rastrigin <- function(x) {
    sum(x^2 - 10 * cos(2 * pi * x)) + 10 * length(x)
}
dimension <- 5
lower <- rep(-5.12, dimension)
upper <- rep(5.12, dimension)
start1 <- rep(1, dimension)
```

Because R handles vectors, this function has as many dimensions as there are parameters in the vector x supplied to it. The traditional function is in two dimensions, but Gubain et al. in **GenSA** suggest that a problem in 30 dimensions is a particularly difficult one. Here we will use five dimensions as a compromise that allows for easier display of results while still providing some level of difficulty. Even the two-dimensional case provides plenty of local minima, as can be seen from the Wikipedia article on the function at http://en.wikipedia.org/wiki/Rastrigin_function. Note that we define lower and upper bounds constraints on our parameters and set a start vector.

15.3.1   Method **SANN** from `optim()`

Base R offers a simulated annealing method in the `optim()` function. As I have indicated, I do not like the fact that this method always performs `maxit` function evaluations.

```
For "SANN" 'maxit' gives the total number of function
evaluations: there is no other stopping criterion. Defaults
to '10000'.
```

Nonetheless, let us try this method on our five-dimensional problem, using all 1's as a start. We will also fix a random number generator seed and reset this for each method to attempt to get some consistency in output.

```
myseed <- 1234
set.seed(myseed)
asann1 <- optim(par = start1, fn = Rastrigin, method = "SANN",
    control = list(trace = 1))

## sann objective function values
## initial        value 5.000000
## iter    1000 value 5.000000
## iter    2000 value 5.000000
## iter    3000 value 5.000000
## iter    4000 value 5.000000
## iter    5000 value 5.000000
## iter    6000 value 5.000000
## iter    7000 value 5.000000
## iter    8000 value 5.000000
## iter    9000 value 5.000000
## iter    9999 value 5.000000
## final          value 5.000000
## sann stopped after 9999 iterations

print(asann1)

## $par
## [1] 1 1 1 1 1
##
## $value
## [1] 5
##
## $counts
## function gradient
##    10000       NA
##
## $convergence
## [1] 0
##
## $message
## NULL
```

Sadly, not only do we make no progress but also the convergence code of 0 implies that we have a "success," even though we have used the maximum number of function evaluations permitted. My opinion is that SANN should be dropped from the function optim().

15.3.2 Package GenSA

Given that **GenSA** uses a 30-dimensional Rastrigin function as its example, we can anticipate a more successful outcome with this tool, and indeed this package does well on this example. Note that we supply only bounds constraints for this method and no starting value.

```
suppressPackageStartupMessages(require(GenSA))
global.min <- 0
tol <- 1e-13
set.seed(myseed)
ctrl <- list(threshold.stop = global.min + tol, verbose = TRUE)
aGenSA1 <- GenSA(lower = lower, upper = upper, fn = Rastrigin,
    control = ctrl)

## Initializing par with random data inside bounds
## It: 1, obj value: 34.82336476
## It: 7, obj value: 9.949580495
## It: 38, obj value: 3.979836228
## It: 63, obj value: 1.989918114
## It: 66, obj value: 0.9949590571
## It: 68, obj value: 7.105427358e-15

print(aGenSA1[c("value", "par", "counts")])

## $value
## [1] 7.10543e-15
##
## $par
## [1]    2.08631e-12 -6.33376e-09  6.58850e-13  1.22032e-13
## [5]    5.46241e-12
##
## $counts
## [1] 1332
```

15.3.3   Packages `DEoptim` and `RcppDE`

The differential evolution approach has been considered by many workers to be one of the more successful tools for global optimization. As with **GenSA**, we supply only the bounds constraints. Our first try stops well before a satisfactory solution has been found and a second try with a larger number of iterations is performed. It should, of course, be possible to use the work already done and recommence the method. However, I have not discovered an obvious and easy way to do this. The same comment applies to several of the other packages discussed in this chapter.

```
suppressPackageStartupMessages(require(DEoptim))
set.seed(myseed)
ctrl <- list(trace = FALSE)
aDEopt1a <- DEoptim(lower = lower, upper = upper, fn = Rastrigin,
    control = ctrl)
print(aDEopt1a$optim)

## $bestmem
##          par1         par2         par3         par4
##   0.005226542 -0.000111016 -0.013303907 -0.017693187
##          par5
## -0.019044041
##
```

```
## $bestval
## [1] 0.174425
##
## $nfeval
## [1] 402
##
## $iter
## [1] 200

tmp <- readline("Try DEoptim with more iterations")

## Try DEoptim with more iterations

set.seed(myseed)
ctrl <- list(itermax = 10000, trace = FALSE)
aDEopt1b <- DEoptim(lower = lower, upper = upper, fn = Rastrigin,
    control = ctrl)
print(aDEopt1b$optim)

## $bestmem
##          par1          par2          par3          par4
##   1.90155e-10   2.39771e-10  -2.15081e-09   5.29262e-10
##          par5
## -3.27813e-10
##
## $bestval
## [1] 0
##
## $nfeval
## [1] 20002
##
## $iter
## [1] 10000
```

RcppDE is possibly more efficient, although on the small example problem, I could not really detect much difference.

```
suppressPackageStartupMessages(require(RcppDE))
set.seed(myseed)
ctrl <- list(trace = FALSE)
aRcppDEopt1a <- DEoptim(lower = lower, upper = upper, fn = Rastrigin,
    control = ctrl)
print(aRcppDEopt1a$optim)

## $bestmem
##          par1          par2          par3          par4          par5
## -0.00563128  -0.00642195  -0.00839810  -0.00115433   0.00311928
##
## $bestval
## [1] 0.0306551
##
## $nfeval
## [1] 10050
##
## $iter
## [1] 200

tmp <- readline("Try RcppDE with more iterations")
```

```
## Try RcppDE with more iterations

set.seed(myseed)
ctrl <- list(itermax = 10000, trace = FALSE)
aRcppDEopt1b <- DEoptim(lower = lower, upper = upper, fn = Rastrigin,
    control = ctrl)
print(aRcppDEopt1b$optim)

## $bestmem
##          par1          par2          par3          par4
## -8.74884e-10 -1.42066e-10 -4.90810e-10  1.65422e-09
##          par5
## -3.18499e-10
##
## $bestval
## [1] 0
##
## $nfeval
## [1] 500050
##
## $iter
## [1] 10000
```

Unfortunately, a simple change in the random number seed gives an unsatisfactory result, even with a lot of computational effort.

```
set.seed(123456)
ctrl <- list(itermax = 10000, trace = FALSE)
aRcppDEopt1b <- DEoptim(lower = lower, upper = upper, fn = Rastrigin,
    control = ctrl)
print(aRcppDEopt1b$optim)

## $bestmem
##          par1          par2          par3          par4
##  5.34667e-10 -1.50326e-09  1.05586e-09 -9.94959e-01
##          par5
##  1.54956e-10
##
## $bestval
## [1] 0.994959
##
## $nfeval
## [1] 500050
##
## $iter
## [1] 10000
```

15.3.4   Package smco

This quite small package (all in R) uses an adaptive coordinate sampling, that is, a one-parameter at a time approach to finding the minimum of a function. It appears to work quite well on the Rastrigin function. The choices of LB and UB instead of lower and upper and trc instead of trace are yet another instance of minor but annoying differences in syntax, although the general structure and output of this package are reasonable.

```
suppressPackageStartupMessages(require(smco))

## This is smco package 0.1

set.seed(myseed)
asmco1 <- smco(par = rep(1, dimension), LB = lower, UB = upper,
    fn = Rastrigin, maxiter = 10000, trc = FALSE)
print(asmco1[c("par", "value")])

## $par
## [1]   1.13780e-04  -4.88992e-06  -1.53800e-04   9.90672e-07
## [5]  -6.44412e-05
##
## $value
## [1] 8.09e-06
```

15.3.5   Package soma

soma, which seems to perform reasonably well on the Rastrigin function, is quite a compact method written entirely in R. Unfortunately, the particular syntax of its input and output make this package somewhat awkward to control. In particular, the documentation does not indicate how to control the output from the method, which turns out to be in package **reportr** by the same developer. This took some digging to discover. Moreover, the names and structures of output quantities are quite different from those in, for example, optim(). **soma** is, however, just a particular example of the problematic diversity of structures and names used in stochastic optimization packages for R.

```
suppressPackageStartupMessages(require(soma))
suppressPackageStartupMessages(require(reportr))
# NOTE: Above was not documented in soma!
setOutputLevel(OL$Warning)
set.seed(myseed)
mybounds = list(min = lower, max = upper)
myopts = list(nMigrations = 100)
asoma1 <- soma(Rastrigin, bounds = mybounds, options = myopts)
# print(asoma1) # Gives too much output -- not obvious to
# interpret.
print(asoma1$population[, asoma1$leader])

## [1]  -3.26797e-08   1.68364e-07  -9.35001e-08  -8.33346e-08
## [5]  -6.61269e-07

print(Rastrigin(asoma1$population[, asoma1$leader]))

## [1] 9.5703e-11
```

15.3.6   Package `Rmalschains`

Quoting from the package DESCRIPTION "Memetic algorithms are hybridizations of genetic algorithms with local search methods." Thus this package is yet another member of a large family of stochastic optimization methods. In contrast to some of the other packages considered here, the output of this package is truly minimal. Unless the `trace` is turned on, it is not easy to determine the computational effort that has gone into obtaining a solution. Furthermore, interpreting the output from the trace is not obvious.

From the package documentation, the package implements a number of search strategies, and is coded in C++. It performs well on the Rastrigin test.

```
suppressPackageStartupMessages(require(Rmalschains))
set.seed(myseed)
amals <- malschains(Rastrigin, lower = lower, upper = upper,
    maxEvals = 10000, seed = myseed, trace = FALSE)
print(amals)

## $fitness
## [1] 7.35746e-09
##
## $sol
## [1]    1.88927e-06   5.70618e-07   1.83268e-07  -4.66274e-06
## [5]    3.37871e-06
```

15.3.7   Package `rgenoud`

The **rgenoud** package is one of the more mature genetic algorithm tools for R. Like others, it has its own peculiarities, in particular, requiring neither a starting vector nor bounds. Instead, it is assumed that the function first argument is a vector, and the `genoud` function simply wants to know the dimension of this via a parameter `nvars`. This package is generally set up for maximization, so one must specify `max=FALSE` to perform a minimization. `print.level` is used to control output.

```
suppressPackageStartupMessages(require(rgenoud))
set.seed(myseed)
agen1 <- genoud(Rastrigin, nvars = dimension, max = FALSE, print.level = 0)
print(agen1)

## $value
## [1] 0
##
## $par
## [1]    3.56310e-14  -1.64324e-13  -2.42290e-12  -1.12555e-12
## [5]   -3.65724e-12
##
```

```
## $gradients
## [1]   0.00000e+00   0.00000e+00 -1.18516e-09 -9.83822e-10
## [5] -2.28939e-09
##
## $generations
## [1] 21
##
## $peakgeneration
## [1] 10
##
## $popsize
## [1] 1000
##
## $operators
## [1] 122 125 125 125 125 126 125 126   0
```

15.3.8   Package GA

This package has many options. Here we will use only a simple setup, which may
be less than ideal. The package is able to use parallel processing, for example.
Because this package uses vocabulary and syntax very different from that in the
optimx family, I have not been drawn to use it except in simple examples to see
that it did work, more or less.

```
suppressPackageStartupMessages(require(GA))
set.seed(myseed)
aGA1 <- ga(type = "real-valued", fitness = function(x) -Rastrigin(x),
    min = lower, max = upper, popSize = 50, maxiter = 1000, monitor = NULL)
print(summary(aGA1))

## +-----------------------------------+
## |            Genetic Algorithm            |
## +-----------------------------------+
##
## GA settings:
## Type                   = real-valued
## Population size        = 50
## Number of generations  = 1000
## Elitism                =
## Crossover probability  = 0.8
## Mutation probability   = 0.1
## Search domain
##          x1     x2     x3     x4     x5
## Min -5.12  -5.12  -5.12 -5.12  -5.12
## Max  5.12   5.12   5.12  5.12   5.12
##
## GA results:
## Iterations             = 1000
## Fitness function value = -4.62579e-05
## Solution               =
##                 x1              x2              x3              x4
## [1,] -6.23299e-05 0.000180146 4.46949e-05 0.000138658
##                 x5
## [1,] -0.00041905
```

15.3.9   Package `gaoptim`

Until I communicated with the maintainer of **gaoptim** (Tenorio, 2013), I had some difficulty in using it, and from the exchange of messages, I suspect the documentation will be clarified. It appears that it is necessary with this package to specify a `selection` method that guides the choice of which trial results are used to drive the algorithm. The simplest choice is "`uniform`" in the emulation of the mating process used to evolve the population of chosen parameter sets.

```
suppressPackageStartupMessages(require(gaoptim))
set.seed(myseed)
minRast <- function(x) {
    -Rastrigin(x)
} # define for minimizing
## Initial calling syntax - - no selection argument
## agaor1<-GAReal(minRast, lb=lower, ub=upper)
agaor1 <- GAReal(minRast, lb = lower, ub = upper, selection = "uniform")
agaor1$evolve(200) # iterate for 200 generations
## The same result was returned from 400 generations
agaor1

## Results for 200 Generations:
## Mean Fitness:
##     Min. 1st Qu.  Median   Mean 3rd Qu.    Max.
##   -88.80   -5.36   -3.13   -8.00   -2.43   -1.42
##
## Best Fitness:
##       Min.  1st Qu.   Median    Mean  3rd Qu.     Max.
##  -43.8000  -1.4300  -0.5420  -1.6900  -0.0045  -0.0045
##
## Best individual:
## [1]   1.10319e-06 -4.53110e-03 -4.48925e-04  1.47530e-03
## [5]   5.03173e-06
##
## Best fitness value:
## [1] -0.00454468
```

We can also select parameter sets based on the "fitness" function, that is, our function being maximized. Unfortunately, interpreting this as a probability is not a good idea for a function we have defined as negative – we are, after all, maximizing this negative function to get near zero for the original function. Rather than have to learn how to write a custom selector function, I have chosen to simply use the option `selector='fitness'`, which uses the objective function values as a proxy for the fitness, along with a redefined objective that is positive. We want larger values as "better." We take the sqrt() to better scale the graph that displays progress (Figure 15.1).

```
maxRast <- function(x) {
    sqrt(1/abs(Rastrigin(x)))
}
agaor2 <- GAReal(maxRast, lb = lower, ub = upper, selection = "fitness")
agaor2$evolve(200)
agaor2
```

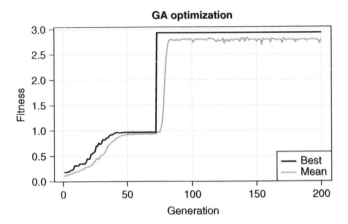

Figure 15.1

```
## Results for 200 Generations:
## Mean Fitness:
##    Min. 1st Qu.  Median    Mean 3rd Qu.    Max.
##   0.108   0.917   2.770   1.950   2.790   2.830
##
## Best Fitness:
##    Min. 1st Qu.  Median    Mean 3rd Qu.    Max.
##   0.182   0.965   2.930   2.130   2.930   2.930
##
## Best individual:
## [1]   1.42045e-02 -1.88966e-02   5.42047e-03 -9.67156e-05
## [5]  -1.38193e-06
##
## Best fitness value:
## [1]  2.92863

plot(agaor2) # note this special method for displaying results

Rastrigin(agaor2$bestIndividual())

## [1]  0.116592
```

Note that, as with most stochastic methods, there are long stretches of work where we do not see any progress.

15.4   Multiple starting values

In many cases, it is possible to simply use a selection of starting vectors with a **local** minimizer to achieve desired results. In the case of the Rastrigin function, this does not appear to be the case. Let us use **Rvmmin** with the bounds specified above and generate 500 random starts uniformly distributed across each coordinate. Note we do not do very well.

```r
suppressPackageStartupMessages(require(Rvmmin))
nrep <- 500
bestval <- .Machine$double.xmax # start with a big number
bestpar <- rep(NA, dimension)
startmat <- matrix(NA, nrow = nrep, ncol = dimension)
set.seed(myseed)
for (i in 1:nrep) {
    for (j in 1:dimension) {
        startmat[i, j] <- lower[j] + (upper[j] - lower[j]) *
            runif(1)
    }
}
for (i in 1:nrep) {
    tstart <- as.vector(startmat[i, ])
    ans <- Rvmmin(tstart, Rastrigin, lower = lower, upper = upper)
    if (ans$value <= bestval) {
        bestval <- ans$value
        cat("Start ", i, " new best =", bestval, "\n")
        bestpar <- ans$par
        bestans <- ans
    }
}
```

```
## Start   1  new best = 34.8234
## Start   3  new best = 14.9244
## Start   7  new best = 9.94958

## Warning: Too many function evaluations

## Start  19  new best = 3.97983

## Warning: Too many function evaluations
## Warning: Too many function evaluations
## Warning: Too many function evaluations

## Start  434  new best = 2.98488
## Start  458  new best = 1.98992
```

```r
print(bestans)
```

```
## $par
## [1] -5.15046e-06  9.94959e-01  1.72078e-07  1.99360e-06
## [5] -9.94959e-01
##
## $value
## [1] 1.98992
##
## $counts
## [1] 98 19
##
## $convergence
## [1] 0
##
## $message
## [1] "Rvmminb appears to have converged"
##
## $bdmsk
## [1] 1 1 1 1 1
```

On the other hand, we do not have to try very hard to find what turns out to be a global minimum of Branin's function (Molga and Smutnicki, 2005), although it is a bit more difficult to find that there are several such minima at different points in the space. Here is the function and its gradient.

```
## branmin.R
myseed <- 123456L # The user can use any seed. BUT ...
some don't work so well.
branin <- function(x) {
    ## Branin's test function
    if (length(x) != 2)
        stop("Wrong dimensionality of parameter vector")
    x1 <- x[1] # limited to [-5, 10]
    x2 <- x[2] # limited to [0, 15]
    a <- 1
    b <- 5.1/(4 * pi * pi)
    c <- 5/pi
    d <- 6
    e <- 10
    f <- 1/(8 * pi)
    ## Global optima of 0.397887 at (-pi, 12.275), (pi, 2.275),
    ## (9.42478, 2.475)
    val <- a * (x2 - b * (x1^2) + c * x1 - d)^2 + e * (1 - f) *
        cos(x1) + e
}
branin.g <- function(x) {
    ## Branin's test function
    if (length(x) != 2)
        stop("Wrong dimensionality of parameter vector")
    x1 <- x[1]
    x2 <- x[2]
    a <- 1
    b <- 5.1/(4 * pi * pi)
    c <- 5/pi
    d <- 6
    e <- 10
    f <- 1/(8 * pi)
    ## Global optima of 0.397887 at (-pi, 12.275), (pi, 2.275),
    ## (9.42478, 2.475) val <- a*(x2 - b*(x1^2) + c*x1 - d)^2 +
    ## e*(1 - f)*cos(x1) + e
    g1 <- 2 * a * (x2 - b * (x1^2) + c * x1 - d) * (-2 * b *
        x1 + c) - e * (1 - f) * sin(x1)
    g2 <- 2 * a * (x2 - b * (x1^2) + c * x1 - d)
    gg <- c(g1, g2)
}
```

In the following, we actually use external knowledge of the three global minima and look at the termination points from different starting points for Rvmmin(). We create a grid of starting points between −5 and 10 for the first parameter and 0 and 15 for the second parameter. We will use 40 points on each axis, for a total of 1600 sets of starting parameters. We can put these on a 40 by 40 character array and assign 0 to any result that is not one of the three known minima, and the labels 1, 2, or 3 are assigned to the appropriate array position of the starting parameters for whichever of the minima are found successfully.

There are very few starts that actually fail to find one of the global minima of this function, and most of those are on a bound. As indicated, the program code labels the three optima as 1, 2, or 3 and puts a 0 on any other solution.

To find out which solution has been generated, if any, let us compute the Euclidean distance between our solution and one of the three known minima, which are labeled y1, y2, and y3. If the distance is less than 1e−6 from the relevant minima, we assign the appropriate code, otherwise it is left at 0.

```
dist2 <- function(va, vb) {
    n1 <- length(va)
    n2 <- length(vb)
    if (n1 != n2)
        stop("Mismatched vectors")
    dd <- 0
    for (i in 1:n1) {
        dd <- dd + (va[i] - vb[i])^2
    }
    dd
}
y1 <- c(-pi, 12.275)
y2 <- c(pi, 2.275)
y3 <- c(9.42478, 2.475)
lo <- c(-5, 0)
up <- c(10, 15)
npt <- 40
grid1 <- ((1:npt) - 1) * (up[1] - lo[1])/(npt - 1) + lo[1]
grid2 <- ((1:npt) - 1) * (up[2] - lo[2])/(npt - 1) + lo[2]
pnts <- expand.grid(grid1, grid2)
names(pnts) = c("x1", "x2")
nrun <- dim(pnts)[1]
```

We will not display the code to do all the tedious work.

```
## Non convergence from the following starts:
##(-3.84615, 0)(6.15385, 0)(6.53846, 0)(5.76923, 0.384615)
##(6.15385, 0.384615)(6.53846, 0.384615)(5.76923, 0.769231)(6.15385, 0.769231)
##(6.53846, 0.769231)(5.76923, 1.15385)(6.15385, 1.15385)(6.53846, 1.15385)
##(5.76923, 1.53846)(6.15385, 1.53846)(5.76923, 1.92308)(6.15385, 1.92308)
##(5.76923, 2.30769)(5.76923, 2.69231)(-1.53846, 3.07692)(-1.15385, 3.46154)
##(-0.769231, 3.84615)(-2.69231, 4.23077)(-2.30769, 4.23077)(-1.53846, 4.23077)
##(-2.30769, 4.61538)(-1.92308, 4.61538)(1.92308, 4.61538)(-1.92308, 5)
##(-0.384615, 5)(0.384615, 5)(1.15385, 5)(1.53846, 5)
##(0.769231, 5.38462)(-0.769231, 5.76923)(-0.384615, 5.76923)(0, 5.76923)
##(0.769231, 5.76923)(-0.769231, 6.15385)(0.384615, 6.15385)(-0.384615, 6.53846)
##(5.76923, 6.53846)(0.384615, 7.30769)(5.76923, 7.69231)(0.769231, 8.07692)
##(1.15385, 8.84615)(1.53846, 10)(6.15385, 11.1538)(6.15385, 11.5385)
##(6.15385, 11.9231)(6.15385, 12.3077)(0.384615, 12.6923)(6.15385, 12.6923)
##(0.384615, 13.0769)(6.15385, 13.0769)(5.76923, 13.4615)(6.15385, 13.4615)
##(6.15385, 13.8462)(6.15385, 14.2308)(0, 14.6154)(5.76923, 14.6154)
##(6.15385, 14.6154)(6.15385, 15)
```

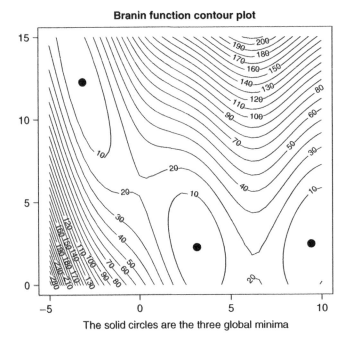

Figure 15.2

We can draw a contour plot of the function and see what is going on (Figure 15.2).

```
zz <- branin(pnts)
zz <- as.numeric(as.matrix(zz))
zy <- matrix(zz, nrow = 40, ncol = 40)
contour(grid1, grid2, zy, nlevels = 25)
points(y1[1], y1[2], pch = 19, cex = 2)
points(y2[1], y2[2], pch = 19, cex = 2)
points(y3[1], y3[2], pch = 19, cex = 2)
title(main = "Branin function contour plot")
title(sub = "The solid circles are the three global minima")
```

Now display the results.

```
for (ii in 1:npt) {
    vrow <- paste(cmat[ii, ], sep = " ", collapse = " ")
    cat(vrow, "\n")
}
```

```
## 1 1 1 1 1 1 1 1 1 1 1 1 1 1 1 2 2 2 2 1 1 2 2 2 2 2 2 2 3 3 0 3 3 3 3 3 3 3 3 3
## 1 1 1 1 1 1 1 1 1 1 1 1 1 1 0 2 2 2 1 1 2 2 2 2 2 2 2 2 3 0 0 3 3 3 3 3 3 3 3 3
## 1 1 1 1 1 1 1 1 1 1 1 1 1 3 2 2 1 1 2 1 2 1 2 2 2 2 2 2 3 0 3 3 3 3 3 3 3 3 3 3
## 1 1 1 1 1 1 1 1 1 1 1 1 1 2 2 2 1 1 1 2 1 2 1 2 2 2 2 2 3 2 0 3 3 3 3 3 3 3 3 3
## 1 1 1 1 1 1 1 1 1 1 1 1 1 2 2 1 1 1 1 1 1 2 2 2 2 2 2 3 0 0 3 3 3 3 3 3 3 3 3 3
## 1 1 1 1 1 1 1 1 1 1 1 1 2 0 1 1 1 1 1 3 2 2 2 2 2 2 3 3 0 3 3 3 3 3 3 3 3 3 3 3
## 1 1 1 1 1 1 1 1 1 1 1 1 1 0 1 1 1 1 2 2 2 2 2 2 2 2 3 3 0 3 3 3 3 3 3 3 3 3 3 3
## 1 1 1 1 1 1 1 1 1 1 1 1 1 1 1 1 1 1 1 1 1 2 2 2 2 2 2 3 3 0 3 3 3 3 3 3 3 3 3 3
## 1 1 1 1 1 1 1 1 1 1 1 1 1 1 1 1 1 1 1 1 1 1 2 2 2 2 2 2 3 2 0 3 3 3 3 3 3 3 3 3
## 1 1 1 1 1 1 1 1 1 1 1 1 1 1 1 1 1 1 1 1 1 2 2 2 2 2 2 2 3 0 3 3 3 3 3 3 3 3 3 3
## 1 1 1 1 1 1 1 1 1 1 1 1 1 1 1 1 1 1 1 1 1 2 2 2 2 2 2 2 2 3 0 3 3 3 3 3 3 3 3 3
## 1 1 1 1 1 1 1 1 1 1 1 1 1 1 1 1 1 1 1 1 1 2 2 2 2 2 2 2 2 3 2 3 3 3 3 3 3 3 3 3
## 1 1 1 1 1 1 1 1 1 1 1 1 1 1 1 1 1 1 1 1 1 2 2 2 2 2 2 2 2 3 2 3 3 3 3 3 3 3 3 3
## 1 1 1 1 1 1 1 1 1 1 1 1 1 1 1 1 1 1 1 0 2 2 2 2 2 2 2 2 3 2 3 2 3 3 3 3 3 3 3 3
## 1 1 1 1 1 1 1 1 1 1 1 1 1 1 1 1 1 1 1 1 2 2 2 2 2 2 2 2 2 3 2 3 3 3 3 3 3 3 3 3
## 1 1 1 1 1 1 1 1 1 1 1 1 1 1 1 1 1 1 0 2 2 2 2 2 2 2 2 2 3 2 3 3 3 3 3 3 3 3 3 3
## 1 1 1 1 1 1 1 1 1 1 1 1 1 1 1 1 1 1 2 2 2 2 2 2 2 2 2 2 3 2 3 3 3 3 3 3 3 3 3 3
## 3 1 1 1 1 1 1 1 1 1 1 1 1 1 1 0 2 2 2 2 2 2 2 2 2 2 2 2 3 2 3 3 3 3 3 3 3 3 3 3
## 3 1 1 1 1 1 1 1 1 1 1 1 1 1 1 2 2 2 3 2 2 2 2 2 2 2 2 0 2 3 3 3 3 3 3 3 3 3 3 3
## 3 3 1 1 1 1 1 1 1 1 1 1 1 1 0 2 2 2 3 2 2 2 2 2 2 2 2 3 2 3 3 3 3 3 3 3 3 3 3 3
## 3 3 1 1 1 1 1 1 1 1 1 1 1 1 2 2 2 2 2 2 2 2 2 2 2 2 2 3 2 3 3 3 3 3 3 3 3 3 3 3
## 3 3 3 1 1 1 1 1 1 1 1 0 1 2 2 2 2 3 2 2 2 2 2 2 2 2 0 2 3 3 3 3 3 3 3 3 3 3 3 3
## 3 3 3 3 1 1 1 1 1 1 0 1 1 0 2 2 2 2 2 2 2 2 2 2 2 2 3 2 3 3 3 3 3 3 3 3 3 3 3 3
## 3 3 3 3 1 1 1 1 1 1 0 0 0 2 0 2 2 2 2 2 2 2 2 2 2 2 3 2 3 3 3 3 3 3 3 3 3 3 3 3
## 3 3 3 3 3 1 1 2 1 1 1 1 2 2 0 2 2 2 2 2 2 2 2 2 2 2 3 2 3 3 3 3 3 3 3 3 3 3 3 3
## 3 3 3 3 3 1 1 2 0 1 1 1 0 2 0 3 0 0 2 2 2 2 2 2 2 2 2 2 3 3 3 3 3 3 3 3 3 3 3 3
## 3 3 3 3 3 3 1 0 0 1 1 1 2 2 2 3 3 3 0 2 2 2 2 2 2 2 2 2 3 3 3 3 3 3 3 3 3 3 3 3
## 3 3 3 3 3 3 0 0 2 0 2 2 3 2 3 3 3 3 3 2 2 2 2 2 2 2 2 2 3 3 3 3 3 3 3 3 3 3 3 3
## 3 3 3 3 3 3 3 2 2 0 3 2 3 2 3 3 3 3 3 3 2 2 2 2 2 2 2 2 3 3 3 3 3 3 3 3 3 3 3 3
## 3 3 3 3 3 3 3 3 2 0 3 3 3 3 3 3 3 3 3 3 2 2 2 2 2 2 2 2 3 3 3 3 3 3 3 3 3 3 3 3
## 3 3 3 3 3 3 3 3 2 3 3 3 3 3 3 3 3 3 3 3 2 2 2 2 2 2 2 0 2 3 3 3 3 3 3 3 3 3 3 3
## 3 3 3 3 3 3 3 3 3 3 3 3 3 3 3 3 3 3 3 3 2 2 2 2 2 2 2 0 2 3 3 3 3 3 3 3 3 3 3 3
## 3 3 3 3 3 3 3 3 3 3 3 3 3 3 3 3 3 3 3 3 2 2 2 2 2 2 2 0 0 3 3 3 3 3 3 3 3 3 3 3
## 3 3 3 3 3 3 3 3 3 3 3 3 3 3 3 3 3 3 3 3 2 2 2 2 2 2 2 0 0 0 3 3 3 3 3 3 3 3 3 3
## 3 3 3 3 3 3 3 3 3 3 3 3 3 3 3 3 3 3 3 3 2 2 2 2 2 2 2 0 0 0 3 3 3 3 3 3 3 3 3 3
## 3 3 3 3 3 3 3 3 3 3 3 3 3 3 3 3 3 3 3 3 2 2 2 2 2 2 2 0 0 3 3 3 3 3 3 3 3 3 3 3
## 3 3 3 0 3 3 3 3 3 3 3 3 3 3 3 3 3 3 3 3 2 2 2 2 2 2 2 0 0 3 3 3 3 3 3 3 3 3 3 3
```

References

Adragni KP, Cook RD, and Wu S 2012 GrassmannOptim: an R package for Grassmann manifold optimization. Journal of Statistical Software 50(5), 1–18.

based on the work of Ivan Zelinka JC 2011 soma: General-purpose optimisation with the Self-Organising Migrating Algorithm. R package version 1.1.0.

Bergmeir C, Molina D, and Benítez JM 2012 Continuous Optimization using Memetic Algorithms with Local Search Chains (MA-LS-Chains) in R. R package version 0.1.

de Colombia PJDVHUN 2012 smco: A simple Monte Carlo optimizer using adaptive coordinate sampling. R package version 0.1.

Eddelbuettel D 2012 RcppDE: Global optimization by differential evolution in C++. R package version 0.1.1. Extends DEoptim which itself is based on DE-Engine (by Rainer Storn).

Mebane WR Jr. and Sekhon JS 2011 Genetic optimization using derivatives: the rgenoud package for R. Journal of Statistical Software 42(11), 1–26.

Molga M and Smutnicki C 2005 Test functions for optimization needs.

Mühlenbein H, Schomisch D, and Born J 1991 The parallel genetic algorithm as function optimizer. Parallel Computing 17(6-7), 619–632.

Mullen K, Ardia D, Gil D, Windover D, and Cline J 2011 DEoptim: an R package for global optimization by differential evolution. Journal of Statistical Software 40(6), 1–26.

Murphy L 2012 Likelihood: Methods for maximum likelihood estimation. R package version 1.5.

Scrucca L 2013 GA: a package for genetic algorithms in R. Journal of Statistical Software 53(4), 1–37.

Tenorio F 2013 Gaoptim: Genetic Algorithm optimization for real-based and permutation-based problems. R package version 1.1.

Törn A and Zilinskas A 1989 Global Optimization, Lecture Notes in Computer Science. New York: Springer-Verlag.

Xiang Y, Gubian S, Suomela B, and Hoeng J 2013 Generalized simulated annealing for global optimization: the gensa package for r. The R Journal 5(1), 13–29.

16

Scaling and reparameterization

Both **scaling** – substituting scalar multiples of optimization parameters for those in the original model – and **reparameterization** – substituting invertible functions of optimization parameters for the parameters themselves – are tools that sometimes assist us to get solutions or else to reduce the computational effort to get solutions. Of course, scaling is but a simple form of reparameterization.

In this chapter, we will look at this topic. Unfortunately, while scaling and reparameterization are often recommended and are mathematically relatively simple, they are not trivial to use because they need to be applied carefully and, as with many topics in nonlinear optimization, the results may both help and impede the process of obtaining good results.

16.1 Why scale or reparameterize?

The essence of nonlinearity of a function $f(\mathbf{x})$ is that the value of $f()$ cannot be expressed in terms where the parameters appear to power 1. This can be important when we want to understand the functional surface of $f()$. In some regions, a standard change – either absolute or percentage – in one of the parameters $\mathbf{x}$ will result in a small change in the value of $f()$ while in others the change will be very large.

From the perspective of optimization, it is helpful to think of the optimization problem as akin to locating the bottom of a bowl. When we can visualize the whole object, this is easy, especially when x is of dimension 2, and where these objects have a size where the width and height are of the same general order of magnitude. When, however, the bowl is 2 mm wide in one direction, 100 m front to back, and a 100 km high, we are dealing with a more difficult task. Similarly, high dimensionality strains our ability to imagine the surface.

Thus our first rationale for **scaling** is that we want to have the sizes of the parameters $\mathbf{x}$ more or less the same magnitude or expressed in units familiar to workers in a field of activity. I find a helpful rule of thumb is that all $x[i]$ for i in

$1 : n$, where n is the number of parameters, should be between 1 and 10. Function minimization methods try to alter the parameters of $\mathbf{x}$ to make the function of the altered parameters $\mathbf{x}_{\text{new}}$ smaller than that for the current set $\mathbf{x}$. Humans find it a great deal easier to envisage changes in the parameters when the parameters and the changes have similar sizes and scales. It may be helpful to think of the parameter sizes as being set on dials like the controls of a stove with settings from 1 to 10. It is very awkward to work with a stove if one burner goes from cold to red hot between 5.01 and 5.03, while another burner has $1.0-9.9$ for the same range of heat.

Worse, there may be regions where it makes no sense to seek an optimum. If the function $f()$ is rapidly oscillating with small changes in one or more of the $\mathbf{x}$, it will be difficult for optimization tools that assume we are "finding the bottom of a bowl" to do their job. Regions where the function is discontinuous or even has discontinuous derivatives can also be problematic. For this discussion, we will set aside these worries, except as they illustrate scaling concerns.

A second rationale is mainly a human issue in managing data and results. In recording or entering data, it is useful to not have to carry too many digits or to have to enter exponents, especially for values of parameters that are "close" to the optimum. For example, in the Hobbs weed infestation problem discussed in the following text, the scaled solution is "near" to the parameter values 2, 5, and 3, which are easily remembered and entered. Moreover, the display of a large number of digits, most of which may have no meaning in the context of the problem at hand, may lead users to believe the results of an estimation procedure are "accurate." The potential for human error increases with the amount of information to be read or copied.

Arguments like this are clearly not sufficient to mandate scaling. Indeed, not all methods need the user to scale the function. However, scaling can be helpful, as we shall try to show by following examples.

A particular issue concerns parameters that are zero at the optimum. For some problems, such zero parameters may not be truly "there." That is, the parameter may not be needed in a model that we are trying to estimate and could be dropped from the function. If the parameter really is part of our objective function, then we may see the scale of the parameter change rapidly relative to other parameters as methods progressively reduce the magnitude of the parameter in an attempt to optimize the objective function.

It is also possible to scale the objective function. I recommend doing this only to make the numbers sensible in the context of the problem or the field of application. Unfortunately, the function `optim()` has a feature for allowing a function scaling to be applied, but its usage example is to change the sign on the function so that the **minimizers** in `optim()` can **maximize** a function. My strong preference is to avoid calling this change of sign a "scaling," especially as maximization appears to be the major, perhaps only, use of the feature.

16.2   Formalities of scaling and reparameterization

Scaling is part of the overall issue of *reparameterization*. For example, if there is an invertible set of functions $\mathbf{z}()$ (we use bold face to indicate that we are using a vector of functions here, one per parameter), we can write

$$\mathbf{x}_{new} = \mathbf{z}(\mathbf{x}) \qquad (16.1)$$

A simple case uses a matrix transformation, for which we use a matrix we will call S for convenience. Thus

$$\mathbf{x}_{scaled} = S\mathbf{x} \qquad (16.2)$$

$$\mathbf{x} = S^{-1}\mathbf{x}_{scaled} \qquad (16.3)$$

While S can be any nonsingular matrix, a simple diagonal matrix with no zero entries on the diagonal provides us with what we often refer to as "scaling," because we then just multiply each element of $\mathbf{x}$ by the corresponding diagonal element of S to get the appropriate element of $\mathbf{x}_{scaled}$. We reverse the transformation by dividing the elements of $\mathbf{x}_{scaled}$ by the diagonal elements of S.

A similarly simple affine transformation uses offsets $\mathbf{o}$, so that the ith reparametrizing function is

$$x_{new_i} = s_i * (x_i - o_i) \qquad (16.4)$$

$$x_i = x_{new_i}/s_i + o_i \qquad (16.5)$$

This is a particular and simple case of an affine transformation between the raw and "scaled" parameters. The so-called "standardized" variables form such transformations using the population means for the shifts and population standard deviations for the scaling. Indeed, base R has a function `scale()` for just this purpose. For most purposes, because such population values are generally unknown, we are better off to use simpler numbers such as powers of 10.

16.3   Hobbs' weed infestation example

We again use the Hobbs problem (Section 1.2).

```
# draw the data
y <- c(5.308, 7.24, 9.638, 12.866, 17.069, 23.192, 31.443, 38.558, 50.156,
    62.948, 75.995, 91.972)
t <- 1:12
plot(t, y)
title(main = "Hobbs' weed infestation data", font.main = 4)
```

The problem was supplied by Mr Dave Hobbs of Agriculture Canada (Figure 16.1). As told to the author, the observations (y) are weed densities per unit area over 12 growing periods. We were never given the actual units of the observations, without which I would now refuse to consider the problem. It was suggested that the appropriate model was a three-parameter logistic, that is,

$$y = b_1/(1 + b_2 \exp(-b_3 t)) \qquad (16.6)$$

where t is the growing period.

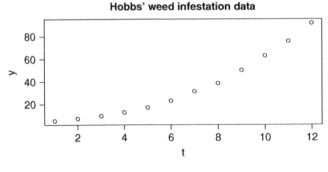

Figure 16.1

Now we will express the sum-of-squares function in R. First we will use an unscaled version of the loss function and residual. Then we will scale by simple powers of 10. Note that the data is embedded in the function.

```
hobbs.f<- function(x){ # # Hobbs weeds problem -- function
    if (abs(12*x[3]) > 50) { # check computability
        fbad<-.Machine$double.xmax
        return(fbad)
    }
    res<-hobbs.res(x)
    f<-sum(res*res)
}
hobbs.res<-function(x){ # Hobbs weeds problem -- residual
# This variant uses looping
    if(length(x) != 3) stop("hobbs.res -- parameter vector n!=3")
    y<-c(5.308, 7.24, 9.638, 12.866, 17.069, 23.192, 31.443,
        38.558, 50.156, 62.948, 75.995, 91.972)
    t<-1:12
    if(abs(12*x[3])>50) {
        res<-rep(Inf,12)
    } else {
        res<-x[1]/(1+x[2]*exp(-x[3]*t)) - y
    }
}
```

Let us try running this problem in `optim()` using the standard Nelder–Mead method starting at the vector expressed as c(100, 10, 0.1). This unscaled start point corresponds to c(1, 1, 1) in the following scaled version.

```
cat("Running the Hobbs problem - code already loaded - with Nelder-Mead\n")

## Running the Hobbs problem - code already loaded - with Nelder-Mead

start <- c(100, 10, 0.1)
f0 <- hobbs.f(start)
cat("initial function value=", f0, "\n")

## initial function value= 10685.3

tu <- system.time(ansnml <- optim(start, hobbs.f,
  control = list(maxit = 5000)))[1]
```

```
# opt2vec is a small routine to convert answers to a vector form for
# printing
Unscaled1 <- opt2vec("unscaled Nelder", ansnm1)

## unscaled Nelder : f( 195.863 , 49.0609 , 0.313756 )= 2.58754
##      after  247  fn &  NA  gr evals
```

The following is a very simple scaling of the original problem. Here is the code for the function and residual.

```
shobbs.f<-function(x){ # # Scaled Hobbs weeds problem -- function
    if (abs(12*x[3]*0.1) > 50) { # check computability
        fbad<-.Machine$double.xmax
        return(fbad)
    }
    res<-shobbs.res(x)
    f<-sum(res*res)
}
shobbs.res<-function(x){ # scaled Hobbs weeds problem -- residual
# This variant uses looping
    if(length(x) != 3) stop("hobbs.res -- parameter vector n!=3")
    y<-c(5.308, 7.24, 9.638, 12.866, 17.069, 23.192, 31.443, 38.558, 50.156,
        62.948, 75.995, 91.972)
    t<-1:12
    res<-100.0*x[1]/(1+x[2]*10.*exp(-0.1*x[3]*t)) - y
}
```

Let us also try running the scaled version of the problem in optim using the standard Nelder–Mead method starting at the vector expressed as c(1, 1, 1).

```
cat("Running the scaled Hobbs problem - code already loaded -
with Nelder-Mead\n")

## Running the scaled Hobbs problem - code already loaded - with Nelder-Mead

start <- c(1, 1, 1)
f0 <- shobbs.f(start)
cat("initial function value=", f0, "\n")

## initial function value= 10685.3

ts <- system.time(ansnms1 <- optim(start, shobbs.f,
  control = list(maxit = 5000)))[1]
opt2vec("scaled Nelder", ansnms1)

## scaled Nelder : f( 1.9645 , 4.9116 , 3.13398 )= 2.58765
##      after  196  fn &  NA  gr evals
```

The examples hint that the scaled version is slightly less work for the Nelder–Mead method to solve but that the advantage is not huge. We may hypothesize

- that scaled problems may have less difficulty finding a solution from a "random" or simple starting point.

- that from equivalent starting points, the scaled function may allow a solution to be found more efficiently, that is, with fewer function evaluations.

Let us apply a different optimizer (BFGS) to the same two tasks.

```
cat("Running the Hobbs problem - code already loaded - with BFGS\n")

## Running the Hobbs problem - code already loaded - with BFGS

start <- c(100, 10, 0.1)
starts <- c(1, 1, 1)
f0 <- hobbs.f(start)
f0s <- shobbs.f(starts)
cat("initial function values: unscaled=", f0, "  scaled=", f0s, "\n")

## initial function values: unscaled= 10685.3    scaled= 10685.3

tu <- system.time(abfgs1 <- optim(start, hobbs.f, method = "BFGS",
 control = list(maxit = 5000)))[1]
opt2vec("BFGS unscaled", abfgs1)

## BFGS unscaled : f( 196.194 ,   49.0884 ,   0.313556 )= 2.58728
##      after   209  fn &   59  gr evals

ts <- system.time(abfgs1s <- optim(starts, shobbs.f, method = "BFGS",
 control = list(maxit = 5000)))[1]
opt2vec("BFGS scaled", abfgs1s)

## BFGS scaled : f( 1.9619 ,   4.90919 ,   3.13568 )= 2.58728
##      after   118  fn &   36  gr evals
```

Here we see that scaling reduces the function/gradient count and gets a slightly lower final function value. Of course, we really should use analytic rather than numerically approximated gradients in the BFGS (actually variable metric) method. Let us put them in and run the same problems again.

```
hobbs.jac <- function(x) {
    # Jacobian of Hobbs weeds problem
    jj <- matrix(0, 12, 3)
    t <- 1:12
    yy <- exp(-x[3] * t)
    zz <- 1/(1 + x[2] * yy)
    jj[t, 1] <- zz
    jj[t, 2] <- -x[1] * zz * zz * yy
    jj[t, 3] <- x[1] * zz * zz * yy * x[2] * t
    return(jj)
}

hobbs.g <- function(x) {
    # gradient of Hobbs weeds problem NOT EFFICIENT TO CALL AGAIN
    jj <- hobbs.jac(x)
    res <- hobbs.res(x)
    gg <- as.vector(2 * t(jj) %*% res)
    return(gg)
}

shobbs.jac <- function(x) {
    # scaled Hobbs weeds problem -- Jacobian
    jj <- matrix(0, 12, 3)
    t <- 1:12
```

```
    yy <- exp(-0.1 * x[3] * t)
    zz <- 100/(1 + 10 * x[2] * yy)
    jj[t, 1] <- zz
    jj[t, 2] <- -0.1 * x[1] * zz * zz * yy
    jj[t, 3] <- 0.01 * x[1] * zz * zz * yy * x[2] * t
    return(jj)
}

shobbs.g <- function(x) {
    # scaled Hobbs weeds problem -- gradient
    shj <- shobbs.jac(x)
    shres <- shobbs.res(x)
    shg <- as.vector(2 * (shres %*% shj))
    return(shg)
}

cat("Running the Hobbs problem - code already loaded - with BFGS\n")

## Running the Hobbs problem - code already loaded - with BFGS

start <- c(100, 10, 0.1)
starts <- c(1, 1, 1)
f0 <- hobbs.f(start)
f0s <- shobbs.f(starts)
cat("initial function values: unscaled=", f0, "  scaled=", f0s, "\n")

## initial function values: unscaled= 10685.3    scaled= 10685.3

tu <- system.time(abfgs1g <- optim(start, hobbs.f, gr = hobbs.g,
 method = "BFGS", control = list(maxit = 5000)))[1]
opt2vec("BFGS unscaled", abfgs1g)

## BFGS unscaled : f( 196.175 ,  49.0885 ,  0.313573 ) = 2.58728
##      after  127  fn &  50  gr evals

ts <- system.time(abfgs1sg <- optim(starts, shobbs.f, gr = shobbs.g,
 method = "BFGS", control = list(maxit = 5000)))[1]
opt2vec("BFGS scaled", abfgs1sg)

## BFGS scaled : f( 1.96186 ,  4.90916 ,  3.1357 ) = 2.58728
##      after  119  fn &  36  gr evals
```

Here we see that we do not gain very much in terms of function/gradient counts by using the analytic gradients for the scaled function, but we do get better results for the unscaled problem, both in terms of function/gradient counts and also a lower final function value. This is likely due to the more or less fixed value used for the "eps" in computing a forward finite-difference approximation to the gradient (Section 10.5). While we actually could alter the shift for each dimension, as the documentation tells us, I do not believe that I have ever done this in R and do not expect users to be able to do so easily.

```
    'ndeps' A vector of step sizes for the finite-difference
        approximation to the gradient, on 'par/parscale' scale.
        Defaults to '1e-3'.
```

Let us try using the `parscale` offered by `optim`. Note that many optimization tools do NOT include this facility. For the BFGS method, we will include a trace so that we can check if the initial function value is as expected. It is easy to get the wrong scaling values in parscale.

```
cat("Try using parscale\n")

## Try using parscale

start <- c(100, 10, 0.1)
anmps1 <- optim(start, hobbs.f, control = list(trace = FALSE,
 parscale = c(100, 10, 0.1)))
opt2vec("Nelder + parscale", anmps1)

## Nelder + parscale : f( 196.45 ,   49.116 ,   0.313398 )= 2.58765
##     after   196  fn &  NA  gr evals

abfps1 <- optim(start, hobbs.f, gr = hobbs.g, method = "BFGS",
 control = list(trace = TRUE, parscale = c(100, 10, 0.1)))

## initial   value 10685.287754
## iter   10 value 1023.938721
## iter   20 value 45.644964
## iter   30 value 2.629944
## final   value 2.587277
## converged

opt2vec("BFGS + parscale", abfps1)

## BFGS + parscale : f( 196.186 ,   49.0916 ,   0.31357 )= 2.58728
##     after   119  fn &  36  gr evals
```

Using `parscale` is, as we see, equivalent within some small rounding errors of using the explicit scaling. Possibly it is my own prejudice, but I prefer the explicit scaling and find it easier to use.

16.4 The KKT conditions and scaling

For an unconstrained objective function $f(\mathbf{x})$, the KKT (Karush–Kuhn–Tucker) conditions (Karush, 1939; Kuhn and Tucker, 1951) for a (local) minimum at $\mathbf{y}$ are that the gradient of f at $\mathbf{y}$ is null and that the eigenvalues of the Hessian of f at $\mathbf{y}$ are all positive. The gradient is the vector of first derivatives of f with respect to the parameters (evaluated at $\mathbf{y}$). The Hessian is the matrix of second derivatives.

Of course, "null" and "positive" are ideas that presume exact arithmetic. The reality is that we want the gradient to be "small." For the Hessian, we want all eigenvalues positive as a first condition. That is relatively easy to check. However, if the ratio of smallest to largest eigenvalue is less than, say, 1e−6, we may suspect that the Hessian is very close to singular. This means that we cannot be sure that we are at a local minimum. Interpreted another way, it suggests that it is difficult to determine the minimum as a region of parameter space may be "flat."

Let us run these checks on the BFGS answer from the scaled Hobbs function using the analytic gradient. The function `kktc()` contains output (in the experimental version used here), so we will avoid duplication of output and not print the result `kbfgslsg`.

```
## Loading required package: numDeriv

## results from KKT check of abfgslsg solution (scaled, with
   gradient from BFGS)
## KKT test tolerances: for g =  0.0012207     for H =  0.012207
## Parameters: [1] 1.96186 4.90916 3.13570
## Successful convergence!
## Compute gradient approximation at finish
## Gradient:[1]   0.001991813 -0.000485584  0.002579320
## kkt1 =   TRUE
## Compute Hessian approximation
##            [,1]       [,2]       [,3]
## [1,]  12654.61 -3256.125 16021.06
## [2,]  -3256.13   862.709 -4095.21
## [3,]  16021.06 -4095.206 20434.34
## Eigenvalues found
## [1] 33860.5130    76.3199    14.8306
## negeig =   FALSE
## evratio =  0.00043799
## kkt2 =   FALSE
```

The result is apparently "not so good." However, if we try the same exercise with the latest parscale version of the unscaled Hobbs function `abfps1` and the unscaled function with gradient (`abfgs1g`), we get the following output.

```
## results from KKT check of abfps1 solution (parscaled, with
   gradient from BFGS)
## KKT test tolerances: for g =  0.0012207     for H =  0.012207
## Parameters: [1] 196.18627  49.09165   0.31357
## Successful convergence!
## Compute gradient approximation at finish
## Gradient:[1]   1.99186e-05 -4.85596e-05  2.57938e-02
## kkt1 =   FALSE
## Compute Hessian approximation
##             [,1]        [,2]        [,3]
## [1,]     1.26546    -3.25613    1602.11
## [2,]    -3.25613     8.62709   -4095.21
## [3,]  1602.10555 -4095.20623 2043434.30
## Eigenvalues found
## [1] 2.04344e+06 4.24926e-01 4.41380e-03
## negeig =   FALSE
## evratio =  2.15998e-09
## kkt2 =   FALSE


## results from KKT check of abfgs1g solution (unscaled, with
   gradient from BFGS)
## KKT test tolerances: for g =  0.0012207     for H =  0.012207
## Parameters: [1] 196.174504  49.088545   0.313573
```

```
## Successful convergence!
## Compute gradient approximation at finish
## Gradient:[1]   3.03187e-05 -7.66194e-04  1.14352e-04
## kkt1 =  TRUE
## Compute Hessian approximation
##            [,1]        [,2]        [,3]
## [1,]    1.26561    -3.25643     1602.15
## [2,]   -3.25643     8.62775    -4095.21
## [3,] 1602.14587 -4095.21284 2043294.41
## Eigenvalues found
## [1] 2.04330e+06 4.24989e-01 4.41682e-03
## negeig = FALSE
## evratio = 2.16161e-09
## kkt2 = FALSE
```

Thus the unscaled functions apparently have better (i.e., smaller) gradients but the Hessian approximations are very close to being effectively singular. This latter feature of the Hobbs problem can cause difficulty for Newton-like methods.

Question: Is the poorer gradient always the case?

Answer: Not necessarily. The following is a very simple two-parameter function, which becomes "badly" scaled as skew gets larger. The solution is at parameter vector $(0, 2)$.

```
diagfn <- function(par, skew = 1) {
    x <- par[1] * skew
    y <- par[2]
    a <- x + y - 2
    b <- x - y + 2
    fval <- a * a + 4 * b * b
}
diaggr <- function(par, skew = 1) {
    x <- par[1] * skew
    y <- par[2]
    a <- x + y - 2
    b <- x - y + 2
    g1 <- (2 * a + 8 * b) * skew
    g2 <- 2 * a - 8 * b
    gr <- c(g1, g2)
    return(gr)
}
```

```
## Diagonal function test
## Nelder-Mead results
##        skew        fmin        par1 par2  nf ng ccode
## an1      1 4.70738e-15 -2.06620e-08    2 105 NA     0
## an1g    10 2.27900e-13  1.57277e-08    2 109 NA     0
## an10   100 4.58041e-12 -5.02533e-09    2 103 NA     0
```

```
## gradient an1:
## [1]   2.28659e-08 -2.58505e-07
## gradient an10:
## [1]  -1.44379e-06  2.57408e-06
```

```
## gradient an100:
## [1] -3.4527e-06  5.7009e-06
##
##  BFGS results
##           skew         fmin           par1 par2 nf ng ccode
## abf1        1 1.77099e-27 -2.21587e-14    2 10  3      0
## abf1g       1 2.83990e-29 -2.84217e-16    2  7  3      0
## abf10      10 1.01635e-19  9.89887e-12    2 26 11      0
## abf10g     10 1.01635e-19  9.89886e-12    2 26 11      0
## abf100    100 3.14335e-25 -1.22168e-15    2 39 10      0
## abf100g   100 3.14640e-25 -1.21920e-15    2 39 10      0


## gradient abf1:
## [1] -1.04805e-13 -6.21725e-14
## gradient abf1g:
## [1] -7.10543e-15  7.10543e-15
## gradient abf10:
## [1] -9.68713e-10  1.72011e-09
## gradient abf10g:
## [1] -9.68713e-10  1.72011e-09
## gradient abf100:
## [1]  1.56142e-12 -2.61657e-12
## gradient abf100g:
## [1]  1.56364e-12 -2.61791e-12
```

If we graph the function for different skew levels, we can see the problem. Here we just show the case for skew=10 (Figure 16.2). Increasing the skew parameter causes the function surface to look less like a bowl and more like a gutter. This is reflected in the differing curvatures in the two dimensions that can also be shown by the eigenvalues of the Hessian at the supposed minimum. However, the gradient size, while greater for the two larger values of skew, does not seem to follow a particular pattern, and what is seen may be rounding error.

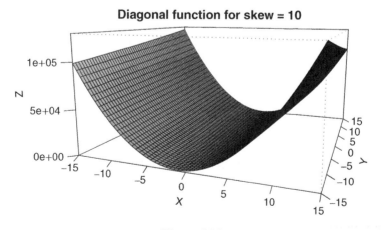

Figure 16.2

```
x = seq(-15, 15, 0.5)
y = seq(-15, 15, 0.5)
## skew=1
nx <- length(x)
ny <- length(y)
z <- matrix(NA, nx, ny)
z10 <- z
z100 <- z
k <- 0
for (i in 1:nx) {
    for (j in 1:ny) {
        par <- c(x[i], y[j])
        val <- diagfn(par, skew = 1)
        val10 <- diagfn(par, skew = 10)
        val100 <- diagfn(par, skew = 100)
        z[i, j] <- val
        z10[i, j] <- val10
        z100[i, j] <- val100
    }
}
persp(x, y, z10, theta = 20, phi = 10, expand = 0.5, col = "lightblue",
  ltheta = 120, shade = 0.75, ticktype = "detailed", xlab = "X", ylab = "Y",
    zlab = "Z")

title("Diagonal function for skew = 10")
```

16.5   Reparameterization of the weeds problem

The three-parameter logistic model used above can take a somewhat different form, namely,

$$y = b_1/(1 + \exp((xmid - t)/scal)) \tag{16.7}$$

where b_1 is often renamed *Asym*. Clearly we can easily transform one form to the other, or to the scaled version shobbs(). The form 16.7 is used in the SSlogis self-starting nonlinear model in R developed by Doug Bates. This form permits a useful geometric interpretation of the parameters in relation to the S-shape of the model. The first parameter, b_1 or *Asym*, is the upper asymptote of the curve; *xmid* is the *t*-axis position of the point where the curve is half way from the lower to the upper asymptote; scal controls the rate at which the curve rises. Smaller values of scal result in curves that rise more steeply. This interpretation is helpful in selecting especially the first two parameters and in guiding the selection of the third. However, does it result in a higher chance of success with our optimizers? We address this after the next section.

16.6   Scale change across the parameter space

Nonlinear optimization is carried out in an Alice-in-Wonderland environment where the scale is different in different parts of the space. To see this more clearly, we can vary the parameters of the different Hobbs models by 1% at the start of the optimization and at its end. For convenience, we work with the scaled Hobbs function as our base model and use the starting parameter vector c(1,1,1). We transform this start for the unscaled Hobbs model and the Hobbs–Bates model. Moreover, we optimize only the scaled Hobbs function and then transform the parameters for the other two models at the supposed optimum. However, we do re-evaluate the unscaled and Bates functions to ensure we have the same solution.

```
##                      fn value 1% p1 effect 1% p2 effect 1% p3 effect
## shobbs - start 10685.28775      -0.941062     0.729787    -0.771187
## uhobbs - start 10685.28775      -0.941062     0.729787    -0.771187
## hbates - start 10685.28775      -0.941062     1.689136    -0.906786
## shobbs - final     2.58765      92.435940    40.769094   387.872740
## uhobbs - final     2.58765      92.435940    40.769094   387.872740
## hbates - final     2.58765      92.435940   605.478530    47.916044
```

The table shows that the effect of shifting one of the parameters by 1% near the starting parameters has a modest effect on the sum-of-squares function value. Indeed, the change in the function value is at most a modest 1.7%. By contrast, at the minimum, changing the asymptote parameter ($b1$, $sb1$, or A_{sym}) changes the sum of squares by better than 92%. For similar percentage changes to the other two parameters, both the hobbs.f() and shobbs.f() change identically, but for a 1% change in the parameters, the functions change roughly 41 and 388%, respectively. The Bates reparameterization is different, with the sum of squares being highly sensitive to the x_{mid} parameter – a 605% change, while a 1% change in *scal* "only" results in a 48% change.

16.7   Robustness of methods to starting points

We are now in a position to examine whether scaling or reparameterization allows us to find solutions more easily. A frequent admonition to novice users of optimization and nonlinear modeling is to choose different and hopefully better starting points for the iterative methods. However, this is not always useful advice. If we knew our answer, we would not need the optimization!

For this exercise, we originally created 1000 points generated pseudo-randomly in a box that is bounded by 0 and 5 for each of the three scaled parameters. We then generated the unscaled and Bates starts. The code looked like this.

```
# generate 1000 sets of points in [0, 5) with runif()
set.seed(12345)
nrun <- 1000
sstart <- matrix(runif(3 * nrun, 0, 5), nrow = nrun, ncol = 3)
ustart <- sstart %*% diag(c(100, 10, 0.1))
bstart <- matrix(NA, nrow = nrun, ncol = 3)
for (i in 1:nrun) {
    b1 <- ustart[[i, 1]]
    b2 <- ustart[[i, 2]]
    b3 <- ustart[[i, 3]]
    Asym <- b1
    scal <- 1/b3
```

```
       xmid <- log(b2)/b3
       bstart[[i, 1]] <- Asym
       bstart[[i, 2]] <- xmid
       bstart[[i, 3]] <- scal
}
```

As the starts are equivalent, we can compare performance for the three models, namely, unscaled, scaled, and Bates. This exercise showed poorer results for the Bates model than I had expected, and in checking, I discovered that some starting parameter sets were inadmissible for the Bates model – infinite or NaN results were returned in function or gradient calculations. I then tried starting with the Bates model and generated points as in the code below. Note that we now restrict the A_{sym} parameter to be larger than 92 (the largest data value). This is not an absolute requirement because data can be higher or lower than our model, but it does avoid some inadmissible starts. Furthermore, the x_{mid} parameter is limited to be between 2 and 11, and scal bigger than 0.05.

```
## random Bates start generate 1000 sets of points in [0, 5) with runif()
set.seed(12345)
nrun <- 1000
bstart <- matrix(runif(3 * nrun, 0, 1), nrow = nrun, ncol = 3)
minasymp <- 92
bstart[, 1] <- bstart[, 1] * (500 - minasymp) + minasymp
bstart[, 2] <- bstart[, 2] * 9 + 2 # inside t interval
scal0 <- 0.05
bstart[, 3] <- bstart[, 3] * (5 - scal0) + scal0 # No check for zero!
sstart <- matrix(NA, nrow = nrun, ncol = 3)
ustart <- sstart
for (i in 1:nrun) {
    Asym <- bstart[[i, 1]]
    xmid <- bstart[[i, 2]]
    scal <- bstart[[i, 3]]
    b1 <- Asym
    b2 <- exp(xmid/scal)
    b3 <- 1/scal
    ustart[[i, 1]] <- b1
    sstart[[i, 1]] <- b1/100
    ustart[[i, 2]] <- b2
    sstart[[i, 2]] <- b2/10
    ustart[[i, 3]] <- b3
    sstart[[i, 3]] <- b3 * 10
}
dput(bstart, file = "supportdocs/scaling/batestrtb.dput")
dput(sstart, file = "supportdocs/scaling/batestrts.dput")
dput(ustart, file = "supportdocs/scaling/batestrtu.dput")
```

However, even with the precautions in generating parameters above for the Bates model, the models sometimes gave infinite or NaN results on initial evaluation when trying the optimizations. Thus we eliminated some points from the starting set by finding the unique set that caused failure for any of the three models.

```r
exlist <- c() # empty
extype <- c() # to record type
nex = 0

cat("Run scaled model -- ")

## Run scaled model --

for (i in 1:nrun) {
    # for (i in 150:160) {
    st <- as.numeric(bstart[i, ])
    ga <- hbates.g(st)
    badgrad <- FALSE
    if (any(is.na(ga)) | any(!is.finite(ga)))
        badgrad <- TRUE
    if (badgrad) {
        exlist <- c(exlist, i)
        extype <- c(extype, "B")
    }
}
cat("found ", length(exlist) - nex, " bad starts\n")

## found  0  bad starts

nex <- length(exlist)

cat("Run scaled model -- ")

## Run scaled model --

for (i in 1:nrun) {
    # for (i in 150:160) {
    st <- as.numeric(sstart[i, ])
    ga <- shobbs.g(st)
    badgrad <- FALSE
    if (any(is.na(ga)) | any(!is.finite(ga)))
        badgrad <- TRUE
    if (badgrad) {
        exlist <- c(exlist, i)
        extype <- c(extype, "S")
    }
}
cat("found ", length(exlist) - nex, " bad starts\n")

## found  0  bad starts

nex <- length(exlist)

cat("Run unscaled model -- ")

## Run unscaled model --

for (i in 1:nrun) {
    # for (i in 150:160) {
    st <- as.numeric(ustart[i, ])
    ga <- hobbs.g(st)
    badgrad <- FALSE
```

```
if (any(is.na(ga)) | any(!is.finite(ga)))
    badgrad <- TRUE
if (badgrad) {
    exlist <- c(exlist, i)
    extype <- c(extype, "U")
}
}
cat("found ", length(exlist) - nex, " bad starts\n")

## found  38  bad starts

nex <- length(exlist)
exlist <- unique(exlist)
options(width = 80)
print(exlist)

##  [1]   11  38  94  99 119 120 175 265 281 347 358 373 442 446 474 538
 541 551 557
## [20] 579 595 599 615 620 623 671 693 723 731 774 776 780 832 871 878
 900 945 986
```

We can now proceed to discover how well different methods fare in optimizing the sum-of-squares function under the three models.

16.7.1   Robustness of optimization techniques

Using package **optimx**, the code for running the calculations is straightforward but tedious and will be omitted here.

```
## Loading required package: optimx

## Result has function value < 2.6
##
##            hbates shobbs uhobbs Total
## ss>=2.6     6155   4860   8101 19116
## ss <2.6     7187   8482   5241 20910
## Total      13342  13342  13342 40026
```

These results indicate that despite the stabilization provided by the reparameterization, the Bates model gives a lower success rate than the scaled Hobbs model, with the unscaled model showing much lower success. If we restrict our measure of success to those results also satisfying the KKT conditions, the unscaled model fails miserably, although this is likely a direct result of the lack of scaling. This lack of scaling increases the condition number of the unscaled Hessian, but the tolerances used for the KKT tests in **optimx** are 0.001 for the gradient and 1e−06 for the ratio of the smallest to largest Hessian eigenvalues. Note that the solution we are working with is the application of optim() method BFGS to the first random start of the unconstrained Hobbs model. We scale this solution and compute the scaled Hobbs model quantities. The application of BFGS to the scaled version of the same start in the scaled Hobbs model actually does not succeed in finding a good solution. Such is life with nonlinear modeling.

```
brunall$lt2.6k <- (brunall$lt2.6 & brunall$kkt1 & brunall$kkt2)
myopt26k <- table(brunall$lt2.6k, brunall$model, useNA = "no")
print(myopt26k)
```

```
##
##          hbates shobbs uhobbs
##   FALSE    7841   7355  13342
##   TRUE     5501   5987      0
```

```
## Get the first unscaled solution parameters
hobbu1 <- as.numeric(c(brunu[1, "b1"], brunu[1, "b2"], brunu[1, "b3"]))
cat("First unscaled solution has sum of squares=", hobbs.f(hobbu1), "\n")
```

```
## First unscaled solution has sum of squares= 22.9947
```

```
require(numDeriv)
hessu1 <- hessian(hobbs.f, hobbu1)
## Hessian of the unscaled function at this solution
print(hessu1)
```

```
##              [,1]          [,2]          [,3]
## [1,]    0.225654    -0.998173       910.864
## [2,]   -0.998173     4.427138     -3989.510
## [3,] 910.863644 -3989.509799 3756349.376
```

```
## Eigenvalues
print(eigen(hessu1)$value)
```

```
## [1]  3.75635e+06  1.94973e-01 -1.96168e-04
```

```
## Gradient of unscaled function
print(as.numeric(hobbs.g(hobbu1)))
```

```
## [1]  0.0751719 -0.1095026 48.5989307
```

```
hobbu1s <- hobbu1 * c(0.01, 0.1, 10) # scale the parameters
## Hessian of the scaled function at this solution (when scaled)
shessu1 <- hessian(shobbs.f, hobbu1s)
print(shessu1)
```

```
##            [,1]      [,2]      [,3]
## [1,] 2256.543  -998.173   9108.64
## [2,] -998.173   442.714  -3989.51
## [3,] 9108.636 -3989.510 37563.49
```

```
## Eigenvalues
print(eigen(shessu1)$value)
```

```
## [1] 40200.24451     63.07288     -0.56663
```

```
## Gradient of scaled function
print(as.numeric(shobbs.g(hobbu1s)))
```

```
## [1]  7.51719 -1.09503  4.85989
```

In the above tables, we are working with 14 methods on slightly fewer than 1000 starting points. We can segment the results by method. We only display the counts where the result is "good," namely, the sum of squares is less than 2.6 and will ignore the KKT conditions.

```
##                hbates shobbs uhobbs    Sum
##
## BFGS              195    298    501    994
## bobyqa            925    842     79   1846
## CG                  8      1      0      9
## hjkb              266    874     19   1159
## L-BFGS-B          944    833    373   2150
## Nelder-Mead       393    675    301   1369
## newuoa            950    848    183   1981
## nlm               186    256    181    623
## nlminb            946    897    861   2704
## nmkb              896    764    588   2248
## Rcgmin            329    358    627   1314
## Rvmmin            210    349    717   1276
## spg                16    754      0    770
## ucminf            923    733    811   2467
## Sum              7187   8482   5241  20910
```

Here we see that the scaled model has a decided advantage over the unscaled one. For some of the methods, the scaled model seems favored in comparison to the Bates reparameterization.

16.7.2   Robustness of nonlinear least squares methods

For nonlinear least squares solvers, we can perform a similar robustness exercise, using the same starting points to allow for comparisons. Here we will not follow through with the Hessian eigenvalues, but only look at results where the sum of squares does not exceed 2.6.

```
## Loading required package: nlmrt
## Loading required package: minpack.lm
```

```
## save computation by reading data
nlruns <- dget(file = "supportdocs/scaling/nlruns.dput")
nlrunu <- dget(file = "supportdocs/scaling/nlrunu.dput")
nlrunb <- dget(file = "supportdocs/scaling/nlrunb.dput")
nlrunall <- rbind(nlruns, nlrunu, nlrunb)
nlrunall <- as.data.frame(nlrunall)
nlrunall[, "meth"] <- as.character(nlrunall[, "meth"])
nlrunall[, "model"] <- as.character(nlrunall[, "model"])
nlrunall <- as.data.frame(nlrunall)
rownames(nlrunall) <- NULL
nlrunall$value[which(is.na(nlrunall$value))] <- 1e+300
nlrunall$lt2.6 <- FALSE
nlrunall$lt2.6[which(nlrunall$value < 2.6)] <- TRUE
## Successes by models and methods
mynltab <- table(nlrunall$lt2.6, nlrunall$meth, nlrunall$model)[2, , ]
mynltab <- addmargins(mynltab)
```

```
rownames(mynltab) <- c(rownames(mynltab)[1:3], "Total")
colnames(mynltab) <- c(colnames(mynltab)[1:3], "Total")
print(mynltab)

##
##          hbates shobbs uhobbs Total
##    nls      902    695    695  2292
##    nlsLM    962    816    817  2595
##    nlxb     960    879    919  2758
##    Total   2824   2390   2431  7645
```

In the above table, it is clear that the Bates form of the model is the easiest for the methods to solve from random starts. It also appears that nlxb() does a little better than nlsLM(). I believe (but have not verified) that this is due to the latter code using some elements of nls() and not attempting a solution when the Jacobian is effectively singular at the start. By contrast, I did not bother to check this possibility in writing nlxb(), and the Levenberg–Marquardt stabilization is able to move the parameters to values where the sum of squares is reduced. I have seen instances of this behavior on several occasions. nls() also seems to exit in a "failure" if a singular Jacobian is encountered along the minimization trajectory.

There are also some "failures" that I did investigate where the reported result is on a plateau or saddle point with small residual and/or near singular Hessian. One such is obtained by nlsLM() or nlxb() from starting point 67 when trying to use the scaled Hobbs model. Note that it is important to properly initiate the random number generator to be able to reproduce results in cases like this. That is, starting parameter set 67 is dependent on how we initiate the random number generator and how we subsequently generate parameter values.

We omit some of the exploratory work to find the particular solution to examine but give a few hints on how the search is done. Here we use the final parameters found by nlsLM() from this starting point with the scaled Hobbs model. Computing the gradient and the Hessian, we see that whether we use scaled or unscaled models, the Hessian is effectively singular, but the gradient is not null. nls() does not return a solution for this start and model, while nlxb() returns a result similar to that examined.

```
which(nlrunall$start == 67)

## [1]  193  194  195 3079 3080 3081 5965 5966 5967

test <- nlrunall[195, ]
print(test)

##          b1       b2      b3   value conv start  model  meth lt2.6
## 195 40.0016 -19420.6 1.80099 9133.98    0    67 shobbs nlsLM FALSE

tpar <- c(test["b1"], test["b2"], test["b3"])
tpar <- unlist(tpar) # this seems to be necessary
spar <- tpar * c(0.01, 0.1, 10) # scaled parameters
names(spar) <- snames
## scaled parameters
```

```
ftest <- shobbs.f(spar)
ftest
```

```
## [1] 9133.98
```

```
gg <- shobbs.g(spar)
gg
```

```
## [1]  0.510063  0.482318 70.795188
```

```
require(numDeriv)
## We use jacobian of gradient rather than hessian of function
hh <- jacobian(shobbs.g, spar)
hh
```

```
##              [,1]        [,2]        [,3]
## [1,] 188371.37634 -2.76132097 -5258.67826
## [2,]     -2.76132  0.00787602     8.03935
## [3,]  -5258.67826  8.03935276  8548.91821
```

```
eigen(hh)$values
```

```
## [1] 1.88525e+05 8.39527e+03 2.90094e-04
```

```
## ... BUT ...
utest <- hobbs.f(tpar)
utest
```

```
## [1] 9133.98
```

```
gu <- hobbs.g(tpar)
gu
```

```
## [1] 5.10063e-03 4.82318e-02 7.07952e+02
```

```
hu <- jacobian(hobbs.g, tpar)
eigen(hu)$values
```

```
## [1] 8.54892e+05 1.85137e+01 2.90094e-06
```

16.8   Strategies for scaling

From the above examples, let us distill some ideas on how to approach scaling in nonlinear models.

First, if your work is exploratory, and you work interactively, it is possible that you do not need to spend a lot of effort on scaling. In this case, my inclination is to try the computations in the form that is presented to you, but watch for obvious differences in scale. If it is easy to adjust the computations, as in the scaled Hobbs versus unscaled form, then that seems to be a worthwhile thing to do.

For scripts that are to be run by others and/or automated into tools where there will be limited chance for meaningful intervention, there is a much higher level of scrutiny needed. One then looks to arrange that parameters are in a reasonable scale and that all have a moderately similar magnitude. Starting values should be considered for their concordance with reality of the underlying system. We did this with the Hobbs problem by considering that the A_{sym} parameter should not be less than the largest value of the modeled variable.

As I will state in other words elsewhere, highly accurate starting values are generally not needed, but really wild or lazy choices are asking for trouble. Generally, some quite simple rules can generate good starts. I recommend loose bounds on parameters to keep them away from ridiculous values.

Transformations are valuable but can be time consuming and error prone to code. However, for parameters that must be positive, I suggest using $blog = \log(b)$ as the working parameter. Suppose we wanted the asymptote $Asym$ to always be positive inside the model functions. Then $Asyml = \log(Asym)$ would be chosen as the working parameter and early in the central function that computes the residuals (for a nonlinear least squares) we would have

```
Asym <- exp(Asyml)
```

$Asym$ is now always positive, but we do not impose a bounds constraint separately.

After results are obtained, I believe it is almost always worth checking the KKT conditions. This can be computationally expensive but making use of erroneous values is much more directly costly to us in money, effort, and reputation. Of course, before running the KKT calculations, it is a really good idea to check that the optimizer used has returned a satisfactory termination code. I have on a number of occasions seen messages "Your program does not work right" from users who have not bothered to note that `conv` or some similarly named return code is telling them the results are likely wrong.

References

Karush W 1939 Minima of functions of several variables with inequalities as side constraints. Master's thesis, MSc Dissertation, Department of Mathematics, University of Chicago, Chicago, IL.

Kuhn HW and Tucker AW 1951 Nonlinear programming. Proceedings of 2nd Berkeley Symposium, pp. 481–492, Berkeley, USA.

17

Finding the right solution

This chapter looks at how we

- find the appropriate solution and

- examining it to decide if it is satisfactory to our purposes.

Almost any developer of software tools for solving computational problems has a few "war stories" about the user who, given a perfectly good mathematical solution to his or her problem, complains that it "is not right." But even before that we have to actually generate a solution.

Throughout this chapter, we will only talk of **minimization** problems. This is the framework we have used elsewhere, and maximization problems are assumed to have been converted to minimizations.

17.1 Particular requirements

As we have touched on earlier, users generally "know" their own problem in ways that the optimization software and its designers cannot be expected to address. If one or more parameters must obey a particular constraint, such as being positive, this **must** be communicated to the software in some way. If a parameter must be an integer, then we need to choose how we solve the problem so that this is taken into account.

The advice in this section, therefore, is

- write down **every** requirement on the solution and solution parameters, even if these conditions may not be imposed explicitly in the software;

- given a trial solution, check that each condition is satisfied;

- as needed, impose the conditions in the software, which may imply a change of method.

Nonlinear Parameter Optimization Using R Tools, First Edition. John C. Nash.
© 2014 John Wiley & Sons, Ltd. Published 2014 by John Wiley & Sons, Ltd.
Companion Website: www.wiley.com/go/nonlinear_parameter

17.1.1   A few integer parameters

In cases where there are at most two or three integer optimization parameters, I would create a loop or grid and solve the optimization problem for each value of the integer in a reasonable range. This is, admittedly, a very crude approach, but mixed integer optimization is a field full of difficulties. Let us look at our classic Rosenbrock function in the integer domain as an example.

```
fr <- function(x) {
    ## Rosenbrock Banana function
    x1 <- x[1]
    x2 <- x[2]
    value <- 100 * (x2 - x1 * x1)^2 + (1 - x1)^2
}
x1r <- seq(-10, 10, 1)
x2r <- x1r

frbest <- 1e+300
xbest <- rep(NA, 2)
for (x1 in x1r) {
    for (x2 in x2r) {
        fval <- fr(c(x1, x2))
        if (fval < frbest) {
            xbest <- c(x1, x2)
            frbest <- fval
        }
    }
}
```

This approach can also be used for situations where we need to decide which sequence of activities is best by generating the set of possible sequences and looping over them. Clearly this is NOT a good approach in general. I would not recommend it, for example, for the type of schedule optimization performed by Michael Trick (see, for example, http://mat.tepper.cmu.edu/trick/). However, when a simple construct such as the example above can be used to perform an essentially exhaustive search, it is likely less work than learning how to set up and use a more sophisticated tool. R is quite weak in this area, and relies on calls to external packages, no doubt because of its statistical heritage. My own experience with integer optimization parameters is severely limited for the same reasons.

17.2   Starting values for iterative methods

Starting values for iterative methods for nonlinear optimization or least squares are a general issue for almost all such software. The choice of starting values is often directed by the purpose of the computations, especially when there are multiple optima in the domain of possible solutions.

From the point of view of testing solutions, starting values may seem an incongruous topic. However, in iterative methods, it is useful to consider not only the parameters that are produced by a method but also where that method started if it used an iterative process. Moreover, if the end point is very far from a start that is considered reasonable, we should wonder what has happened. The test, if we can consider it as such, is one of consistency with our anticipated outcome.

The starting parameters we use come from a variety of sources:

- For a number of test problems, there are well-known sets of starting values. Hence, for the Rosenbrock test function (Rosenbrock, 1960), the start is at $(-1.2, 1)$.

- For some problems, such as the `nls()` problems with `selfStart()` modules, there are starting values that satisfy some supposedly "reasonable" conditions. Note the capitalization in `selfStart`.

- For problems with which the user has a base of experience, it is likely that quite good guesses at the parameters can be provided.

- If there are sensible bounds on the parameters, then some choice such as the midpoints of the intervals may work reasonably well. For positive parameters, the square root of the upper bound may be a better choice (i.e., the geometric mean of the bounds).

- Where possible, I recommend that starting values are NOT on a bound. Some methods (e.g., `nmkb()` in **dfoptim**) cannot be called with parameters on the boundary.

- For problems that are well known to be reliably solved from practically any starts, it is not uncommon to use 1 for every starting value. This is, in fact, used as a default by `nls()` and some other programs, and I have on occasion used it myself.

In many cases, very crude approximations will suffice to get the scale of parameters, and some commonsense thinking will provide loose bounds. Better starting values can be obtained if a rough linearization of a model is possible so that, for example, a linear regression can be performed. The starting values for the `SSlogis()` tool use a variant of this idea (see Sections 6.5.1 and 16.5). Note that it is possible to display the code to see what this does by typing `SSlogis` at the R command line.

17.3   KKT conditions

17.3.1   Unconstrained problems

For unconstrained minimization of functions that have continuous gradients, the solution must have a zero gradient at every local minimum (Karush, 1939; Kuhn and Tucker, 1951). This is a necessary condition, but because local maxima and saddle points also give zero gradients, it is not sufficient. We call this the first-order KKT (Karush–Kuhn–Tucker) condition.

We also want the objective function to be higher at every point "near" the proposed minimum, so that the function is locally convex. There is a nice mnemonic at http://shreevatsa.wordpress.com/2009/03/17/convex-and-concave/ to help one remember this. If the second-order derivatives of the objective can be computed, they define the Hessian matrix

$$H_{i,j} = \partial^2 f / \partial x_i \partial x_j$$

The condition for the proposed solution to be at "the bottom of a bowl" is that the Hessian is positive-definite, for which the most common test is that its eigenvalues are all positive.

Reality, of course, is never so nice. The villain in this is the possibility of floating-point representations of eigenvalues that are "small." Indeed, we can likely create matrices that resemble Hessians that are provably positivee-definite in exact arithmetic that give some small negative estimates of the eigenvalues. And it is very common that real problems give positive but small Hessian eigenvalues. Thus we must test for "small."

In this activity, my own rule of thumb is to first look for all positive eigenvalues. Those computed negative or zero should, I believe, be reported as such, even if it turns out later that they are an awkward special case of unfortunate rounding. In any event, these cases are candidates for careful examination and will likely cause trouble for our optimizers.

Assuming all the eigenvalues are strictly positive, I take the ratio of the smallest to the largest of the Hessian eigenvalues. This ratio should not be "too small." In package **optimx**, ratios smaller than 1e−6 are taken as evidence of effective singularity of the Hessian.

There is no absolute rule about such tolerances. My own justification comes from multiple linear regression, where the singular values of the matrix of variables are comparable to the square roots of the Hessian eigenvalues. Thus a ratio of singular values smaller than 1e−3 would be taken as evidence of singularity, which in linear regression is indicative of a lack of independence between the explanatory variables. The singular values give the relative contribution of orthogonalized explanatory variables to the overall variability of these variables. Thus a singular value that is 1/1000 as big as the largest is contributing a very tiny fraction of the overall variability. Indeed, if we graph the singular values as bars on a regular-sized sheet of paper, the smallest singular value is represented by a bar about 0.2 mm long or less, because A4 and "letter" sizes are 210 and 216 mm wide, respectively.

Warning: Many people assume that computed eigenvalues will be sorted largest to smallest. While many programs do indeed provide them in this ordering, it pays to be careful and check or else use program code that explicitly looks for the maximal and minimal values. The extra time for the check is small compared to the time to repair the damage when we get this wrong.

17.3.2 Constrained problems

The optimality conditions for constrained problems are more demanding to both specify and compute. Gill et al. (1981, Chapter 3) give a fairly comprehensive solution, but for most R users, I believe that the material will not be easy reading. There is a good summary introduction to the wider topic in http://en.wikipedia. org/wiki/Karush%E2%80%93Kuhn%E2%80%93Tucker_conditions.

Clearly, if the constraints are not active, we can use the KKT conditions above. When one or more constraints are active, however, we need to ensure that the gradient as projected along any feasible direction is essentially zero as a first-order condition. There are extensions of the KKT conditions for such situations, but we need to be conscious that they depend on assumptions of continuity and differentiability that may not be true.

My own preference – and this is colored by the limited number of problems with general constraints I have encountered – is to test the proposed optimum to see if it satisfies the constraints. This has the merit that the computations are done using the user's own code. We can also check nearby points to ensure that they are either infeasible (we are on a constraint boundary) or have a higher value of the objective function.

17.4   Search tests

Following the idea of checking points near a proposed solution, I have found it useful for a number of problems to perform a simple axial search around a proposed or trial minimum. Here we will consider only **minimization**. I have found in my own work that trying to add maximization in a clean way to such search strategies – which the user may wish to alter – is a recipe for trouble. For maximization problems, we minimize (−1) times the function and report the negated value. (It is the process of trying to apply scalings appropriately for reporting results that creates the issues for searches.)

The most common search I apply is an axial search where along each parameter direction a small step "forward" and "backward" is attempted and the objective function reevaluated. Code for this in BASIC appeared in Nash and Walker-Smith (1987). There is also a version in R in the package **optextras** in the **optimizer** project packages on R-forge (https://r-forge.r-project.org/R/?group_id=395). This is able to recognize bounds and mask constraints. However, it does not evaluate more general constraints and would need modification for them.

One of the purposes for which the axial search was implemented was to restart direct search algorithms like Nelder–Mead. It is, of course, not foolproof. For example, a function having a saddle between peaks to the northwest and southeast of a supposed minimum that falls away in the northeast and southwest directions could find that the north, south, east, and west axial directions all give higher points on our fog-bound surface.

References

Gill PE, Murray W, and Wright MH 1981 Practical Optimization. Academic Press, London and New York.

Karush W 1939 Minima of functions of several variables with inequalities as side constraints. Master's thesis, M.Sc. Dissertation, Department of Mathematics, University of Chicago Chicago, IL.

Kuhn HW and Tucker AW 1951 Nonlinear programming. Proceedings of 2nd Berkeley Symposium, pp. 481–492, Berkeley, USA.

Nash JC and Walker-Smith M 1987 Nonlinear Parameter Estimation: An Integrated System in BASIC. Marcel Dekker, New York. See http://www.nashinfo.com/nlpe.htm for an expanded downloadable version.

Rosenbrock HH 1960 An automatic method for finding the greatest or least value of a function. The Computer Journal 3, 175–184.

18

Tuning and terminating methods

This chapter is about improving the performance of computational processes. For most calculations, I do not find it necessary to bother with measuring how fast my work is getting done. However, there are situations where a calculation must be performed many times and it is also slow enough to complete that needed results are not available quickly enough.

18.1 Timing and profiling

R offers the function `system.time()` for timing commands. In particular, we usually select the third element of the returned vector as the time taken by the command to execute. Here is an example where we time how long it takes to generate a set of random uniform numbers.

```
nnum <- 100000L
cat("Time to generate ", nnum, " is ")

## Time to generate  100000   is

tt <- system.time(xx <- runif(nnum))[[3]]
cat(tt, "\n")

## 0.005
```

Besides the `system.time` facility, R offers some packages to assist in timing. Unfortunately, while the English word "benchmark" would be a logical choice, both "benchmark" and "benchmarking" refer to packages for other

Nonlinear Parameter Optimization Using R Tools, First Edition. John C. Nash.
© 2014 John Wiley & Sons, Ltd. Published 2014 by John Wiley & Sons, Ltd.
Companion Website: www.wiley.com/go/nonlinear_parameter

calculations. **benchmark** is about statistical learning methods (Eugster and Leisch, 2011; Eugster et al. 2012), and while it may have some features useful for timing R expressions, there are many of these features and most are not directed to the task at hand. Thus **benchmark** is **not** one of the tools I use. Similarly, package **Benchmarking** is about stochastic frontier analysis, a form of economic modeling.

Two packages that **are** useful are **rbenchmark** and **microbenchmark**. R also has a profiling tool Rprof().

18.1.1   rbenchmark

Within this package (Kusnierczyk, 2012), the function benchmark is a wrapper for system.time to allow for timing a number of replications of a command or set of commands. Here we will content ourselves with showing the simplest use of this tool. Note that we refer explicitly to rbenchmark::benchmark() to avoid an unfortunate name conflict with package **Rsolnp**, which is present when the text for this book is processed by the **knitr** package.

```
nnum <- 100000L
require("rbenchmark")

## Loading required package: rbenchmark
cat("Time to generate ", nnum, " is ")

## Time to generate  100000   is

trb <- rbenchmark::benchmark(runif(nnum))
# print(str(trb))
print(trb)

##          test replications elapsed relative user.self sys.self user.child
## 1 runif(nnum)          100   0.502        1       0.5    0.004          0
##    sys.child
## 1          0
```

18.1.2   microbenchmark

Of the timing tools, I find that I use this the most. It gives a five-number summary of the distribution of the times over the number of replications.

```
nnum <- 100000L
require("microbenchmark")

## Loading required package: microbenchmark

cat("Time to generate ", nnum, " is ")

## Time to generate  100000   is

tmb <- microbenchmark(xx <- runif(nnum))
print(str(tmb))
```

```
## Classes 'microbenchmark' and 'data.frame': 100 obs. of  2 variables:
##  $ expr: Factor w/ 1 level "xx <- runif(nnum)": 1 1 1 1 1 1 1 1 1 1 ...
##  $ time: num  4603772 4607334 4601667 4605348 4597553 ...
## NULL

print(tmb)

## Unit: milliseconds
##                expr     min      lq median      uq     max neval
##  xx <- runif(nnum) 4.59674 4.60669 5.12193 5.12774 7.39667   100
```

microbenchmark (Mersmann, 2013) also offers a boxplot facility to graph the results. For both this and **rbenchmark**, multiple statements may be timed and there are a number of other controls. I also find that I often take the time element of the returned object, thus tmb$time in the example above, and compute its mean and standard deviation.

18.1.3   Calibrating our timings

Comparing timings is not always easy. First, it will be silly to compare machines with different underlying capabilities. Thus we want to make our measurements on a single machine. Second, we want to have an idea of how fast that machine is. There is a long history of such benchmark programs, for example, Dongarra et al. (2003), and plenty of examples such as in http://www.netlib.org/benchmark/. In the present situation, I find it helpful to have a small program actually in R with which to calibrate a particular machine. Here is my choice. Because this takes a while to run, we perform the computation off-line and save the results.

```
### Compute benchmark for machine in R J C Nash 2013
busyfnloop <- function(k) {
    for (j in 1:k) {
        x <- log(exp(sin(cos(j * 1.11))))
    }
    x
}
busyfnvec <- function(k) {
    vk <- 1:k
    x <- log(exp(sin(cos(1.1 * vk))))[k]
}
require(microbenchmark, quietly = TRUE)
nlist <- c(1000, 10000, 50000, 1e+05, 2e+05)
mlistl <- rep(NA, length(nlist))
slistl <- mlistl
mlistv <- mlistl
slistv <- mlistl
for (kk in 1:length(nlist)) {
    k <- nlist[kk]
    cat("number of loops =", k, "\n")
    cat("loop\n")
    tt <- microbenchmark(busyfnloop(k))
```

```
    mlistl[kk] <- mean(tt$time) * 1e-06
    slistl[kk] <- sd(tt$time) * 1e-06
    cat("vec\n")
    tt <- microbenchmark(busyfnvec(k))
    mlistv[kk] <- mean(tt$time) * 1e-06
    slistv[kk] <- sd(tt$time) * 1e-06
}
## build table
cvl <- slistl/mlistl
cvv <- slistv/mlistv
l2v <- mlistl/mlistv
restime <- data.frame(nlist, mlistl, slistl, cvl, mlistv, slistv, cvv, l2v)
names(restime) <- c(" n", "mean loop", "sd loop", "cv loop", "mean vec",
"sd vec", "cv vec", "loop/vec")
## Times were converted to milliseconds
restime
## dput(restime, file='filenameforyourmachine.dput')
```

```
##       n mean loop  sd loop   cv loop  mean vec   sd vec   cv vec loop/vec
## 1 1e+03   1.73955 0.227875 0.1309960  0.183392 0.073123 0.398724  9.48542
## 2 1e+04  17.35379 0.461499 0.0265936  2.019939 2.138447 1.058669  8.59124
## 3 5e+04  87.37063 3.499007 0.0400479  9.592821 2.801119 0.292002  9.10792
## 4 1e+05 175.11408 4.565432 0.0260712 19.563233 5.574316 0.284938  8.95118
## 5 2e+05 350.09785 6.398471 0.0182762 38.632340 7.214505 0.186748  9.06230
```

For this book, I have been running on the so-called "desktop" computer (although mine is in a tower case and sits on the floor) that has a "AMD Phenom(tm) II X6 1090T Processor" as revealed by typing `cat /proc/cpuinfo` in a terminal window, rather than trusting the purchase invoice. This processor has six cores and runs at 800 MHz with a 64 bit instruction set. There is a 16 GB RAM in the box, confirmed with `cat/proc/meminfo` However, all this information is somewhat disconnected from what R can actually use of the machine's capabilities. In the example machine, which I call J6, R should be able to access almost all of the available memory. On the other hand, the speed with which it is able to do so depends on the motherboard and its circuitry to support the processor and memory, as well as software at the operating system, compiler, and R levels. Changes in performance can be expected when the changes occur to the operating and computational software. These occasionally get noted on the R mailing lists (e.g., http://r.789695.n4.nabble.com/big-speed-difference-in-source-btw-R-2-15-2-and-R-3-0-2-td4679314.html). Hence the utility of a small and relatively quick program to calibrate our timings.

Note that these timings will not be reproducible. Modern computers are performing many tasks, and these interfere with each other to some extent. To see this, simply run the program above several times and compare the results. You can even save the dataframe. Note that this variation is **after** we have averaged 100 executions of each case, yet I regularly observe differences in mean times of 2–3% percent.

18.2   Profiling

Profiling is a generally quite powerful technique to time a program by its component pieces. The purpose is to discover where the computations take time. Generally, it is NOT an easy job, because the act of capturing the time needed to do something actually takes some time itself, and the measurement interval may, on modern machines, be large enough that we finish the computation before we can measure it.

Thus profiling is best used on calculations we have already discovered are "slow." The following example shows how it is done by wrapping a call to the Rvmmin minimizer in commands that ask for a profile to be saved. Then we recall and crudely analyze this data before displaying it.

The following code put in file `rpcode.R` will be profiled.

```
require(Rvmmin)
sqs<-function(x) {
   sum( seq_along(x)*(x - 0.5*seq_along(x))^2)
}
sqsg<-function(x) {
   ii<-seq_along(x)
   g<-2.*ii*(x - 0.5*ii)
}
xstrt<-rep(pi,200)
aa<-0.5*seq_along(xstrt)
ans<-Rvmmin(xstrt, sqs, gr=sqsg)
```

We will run this with code and put it in file `doprof.R`, that calls the profiler. This clumsy approach is necessary here because it does not seem possible to do the profiling inside the **knitr** markdown environment.

```
tryit<-Rprof("testjohn1.txt", memory.profiling=TRUE)
source("rpcode.R", echo=TRUE)
Rprof(NULL)
```

We actually put everything together as follows:

```
system("Rscript --vanilla doprof.R")
myprof <- summaryRprof("testjohn1.txt")
myprof$by.self

##               self.time  self.pct  total.time  total.pct
## "%*%"              0.42     35.59        0.42      35.59
## "*"                0.24     20.34        0.24      20.34
## "-"                0.20     16.95        0.20      16.95
## "crossprod"        0.08      6.78        0.08       6.78
## "Rvmminu"          0.06      5.08        1.08      91.53
## "+"                0.04      3.39        0.04       3.39
## "seq"              0.04      3.39        0.04       3.39
## "as.vector"        0.02      1.69        0.32      27.12
## "func"             0.02      1.69        0.06       5.08
## "grepl"            0.02      1.69        0.02       1.69
## "objects"          0.02      1.69        0.02       1.69
## "t"                0.02      1.69        0.02       1.69
```

In this example, it can be noted that the `outer()` function, which is used in Rvmmin to perform the BFGS update, is a relatively large user of the overall computing time. Can we do better? Indeed we can, but as with most performance matters, there are many tedious details.

Another profiling tool is Hadly Wickham's **profr**, which generates data into a structure very similar to a data frame. Again, it does not seem possible to run this inside **knitr**. There is a plot method for **profr**, which presents the timing information in a relatively interesting way, although I did not find it immediately useful. As with all profiling tools, some effort is required to focus on the pieces of code that are worth attention.

```
require(profr)
tryit<-profr("rpcode.R")
tryit
```

18.2.1 Trying possible improvements

On the basis of the profile mentioned earlier, potential improvements to the BFGS update of the approximate inverse Hessian in the program Rvmminu() need to be tested for correctness as well as performance. The essential computation updates a matrix B by applying outer products of vectors c and t which are derived from available search direction and gradient information. Our first step was to use the R command dput() to save the contents of examples of the two vectors and the matrix to files testc.txt, "testt.txt," and "testB.txt," which we then use in our measurements.

The measurements are made using microbenchmark(). The four variants of the update were the original, coded in bfgsout, a form using simple loops in bfgsloop, a matrix multiplication form in bfgsmult, and, finally, an outer product variant where we save an intermediate result in a second matrix (bfgsx). The last variant costs us the storage of a matrix, which is potentially a large bite of available memory if the number of parameters to optimize is large.

We also look at speeding up some other parts of the computation, namely, the calculation of the quantities D1 and D2 and the building of an intermediate vector y.

```
require(microbenchmark)
# Different ways to do the BFGS update
bfgsout <- function(B, t, y, D1, D2) {
    Bouter <- B - (outer(t, y) + outer(y, t) - D2 * outer(t, t))/D1
}

bfgsloop <- function(B, t, y, D1, D2) {
    A <- B
    n <- dim(A)[1]
    for (i in 1:n) {
        for (j in 1:n) {
            A[i, j] <- A[i, j] - (t[i] * y[j] + y[i] * t[j] - D2 * t[i] *
t[j])/D1
        }
    }
    A
```

```
}

bfgsmult <- function(B, t, y, D1, D2) {
    A <- B
    A <- A - (t %*% t(y) + y %*% t(t) - D2 * t %*% t(t))/D1
}

bfgsx <- function(B, t, y, D1, D2) {
    # fastest
    A <- outer(t, y)
    A <- B - (A + t(A) - D2 * outer(t, t))/D1
}

# Recover saved data
c <- dget(file = "supportdocs/tuning/testc.txt")
t <- dget(file = "supportdocs/tuning/testt.txt")
B <- dget(file = "supportdocs/tuning/testB.txt")
# ychk<-dget(file='supportdocs/tuning/testy.txt')
tD1sum <- microbenchmark(D1a <- sum(t * c))$time # faster?
tD1crossprod <- microbenchmark(D1c <- as.numeric(crossprod(t, c)))$time
# summary(tD1sum) summary(tD1crossprod)
mean(tD1crossprod - tD1sum)

## [1] 697.22

cat("abs(D1a-D1c)=", abs(D1a - D1c), "\n")

## abs(D1a-D1c)= 0

if (D1a <= 0) stop("D1 <= 0")
tymmult <- microbenchmark(y <- as.vector(B %*% c))$time
tycrossprod <- microbenchmark(ya <- crossprod(B, c))$time # faster
# summary(tymmult) summary(tycrossprod)
mean(tymmult - tycrossprod)

## [1] 20977.2

cat("max(abs(y-ya)):", max(abs(y - ya)), "\n")

## max(abs(y-ya)): 0

td2vmult <- microbenchmark(D2a <- as.double(1 + (t(c) %*% y)/D1a))$time
td2crossprod <- microbenchmark(D2b <- as.double(1 + crossprod
 (c, y)/D1a))$time # faster
cat("abs(D2a-D2b)=", abs(D2a - D2b), "\n")

## abs(D2a-D2b)= 0

# summary(td2vmult) summary(td2crossprod)
mean(td2vmult - td2crossprod)

## [1] 11249.4

touter <- microbenchmark(Bouter <- bfgsout(B, t, y, D1a, D2a))$time
tbloop <- microbenchmark(Bloop <- bfgsloop(B, t, y, D1a, D2a))$time
tbmult <- microbenchmark(Bmult <- bfgsloop(B, t, y, D1a, D2a))$time
tboutsave <- microbenchmark(Boutsave <- bfgsx(B, t, y, D1a, D2a))$time
# faster?
```

```
## summary(touter) summary(tbloop) summary(tbmult) summary(tboutsave)
dfBup <- data.frame(touter, tbloop, tbmult, tboutsave)
colnames(dfBup) <- c("outer", "loops", "matmult", "out+save")
boxplot(dfBup)
title("Comparing timings for BFGS update")


## check computations are equivalent
max(abs(Bloop - Bouter))

## [1] 1.38778e-17

max(abs(Bmult - Bouter))

## [1] 1.38778e-17

max(abs(Boutsave - Bouter))

## [1] 0
```

From these results, it appears that using the outer product function in R is already the best choice for performance in computing the update of the approximate inverse Hessian that is critical to a number of quasi-Newton and variable metric methods. For computing the D1 quantity that enters this update, the sum(t*c) mechanism is slightly better than using crossprod(t,c). We can also make some modest improvements by using crossprod() to compute the intermediate vector y and the quantity D2. These improvements will percolate into the distributed code.

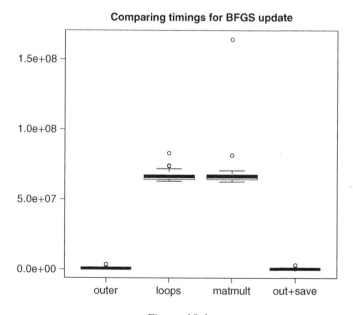

Figure 18.1

18.3   More speedups of R computations

We have just seen one example of how we might improve the speed of a function by substituting equivalent code. We will return to this shortly, but first will consider the byte-code compiler for R.

18.3.1   Byte-code compiled functions

One of the easiest tools for speeding up R is the so-called byte-code compiler due to Luke Tierney, which has been part of the standard distribution since version 2.14. There is also just-in-time compilation in the Ra system due to Stephen Milborrow (http://www.milbo.users.sonic.net/ra/), but I so far have no experience with that.

The treatment here will be limited to the use of the cmpfun() function, but readers who must squeeze every unnecessary microsecond out of their code should read the manual and run timing experiments. Furthermore, it is important to note that this facility can be very helpful, but my experience, echoing comments from Tierney and others, is that code using vectorized commands and functions like crossprod() or involving unsubscripted vectors will see little or no improvement. Code with vector element operations generally benefits quite substantially.

The compiler package is part of the distribution but needs to be loaded or invoked via appropriate instructions when packages are installed. Here we will use it explicitly. The compiler can be loaded, and the manual displayed, by R commands

```
require(compiler)
?compile
```

Note that it is possible to use this implicitly, so that functions are compiled as loaded. Sometimes the byte-code compiler is referred to as a "just-in-time" compiler, because R can, with appropriate options set, apply the compiler to code that is loaded via the source() command. My preference is to run things explicitly.

18.3.2   Avoiding loops

In R, the folklore is that loops are particularly "slow." Moreover, while loops are considered slower than for loops. We can check that suggestion and in the process demonstrate the byte-code compilation possibility.

```
# forwhiletime.R
require(microbenchmark)
require(compiler)

tfor <- function(n) {
    for (i in 1:n) {
        xx <- exp(sin(cos(as.double(i))))
    }
    xx
}
```

```
twhile <- function(n) {
    i <- 0
    while (i < n) {
        i <- i + 1
        xx <- exp(sin(cos(as.double(i))))
    }
    xx
}
n <- 10000

timfor <- microbenchmark(tfor(n))
timwhile <- microbenchmark(twhile(n))
tforc <- cmpfun(tfor)
twhilec <- cmpfun(twhile)
timforc <- microbenchmark(tforc(n))
timwhilec <- microbenchmark(twhilec(n))
looptimes <- data.frame(timfor$time, timforc$time, timwhile$time,
timwhilec$time)
colMeans(looptimes)
```

```
##     timfor.time    timforc.time   timwhile.time timwhilec.time
##        12883398        9177664        19394950       10354892
```

We observe that `for` loops take less than two-third the time of `while` loops in interpreted code but are only about 10% faster when that code is compiled. Moreover, the `while` loops gain a lot from compilation.

As indicated, R does best at computations that involve vectors, so that loops are not used explicitly. That is, when we can write an operation so that whole vectors undergo the same process, R generally performs well. Thus, `sum(x * y)` where `x` and `y` are vectors is fast. Similarly, `crossprod()` is efficient (but outputs a matrix structure). Loops or matrix multiplications (using `%*%`) are much slower. If we need to speed things up, it is important to run timings and to do so in the computing environment we will actually be using. The examples in this chapter are illustrations.

18.3.3   Package upgrades - an example

A general mantra among those who advise others about R is to use the latest versions of R and of packages. Owing to a rather unusual computational model, R can sometimes perform some operations very efficiently (from the point of view of machine cycles), but others can have unusually long execution times. For example, in 2011, I commented that I had two variants of a differential equation model I was trying to estimate by maximum likelihood, and one of these took almost 800 times as long to compute as the other. See the original post and follow the thread from http://tolstoy.newcastle.edu.au/R/e15/devel/11/08/0373.html. Such differences, when both codes are "reasonable" to an initial view, can be very annoying for users and can lead to criticism of R that is incorrect in a global sense, although clearly valid for the particular examples.

As I was about to put this example here, I had occasion to review the e-mail discussion referenced earlier. As I did a Google search, I also found a slightly later contribution by Martin Maechler, noting a new version of package **expm**. There are also more recent versions of package **Matrix**. Using the latest CRAN offerings in July 2013, I was not able to reproduce the extreme differences in timing that were observed in 2011. The data and the code to compute the two negative log likelihoods are as follows, along with some checks on the results obtained. Note that we execute this code off-line on a particular machine (J6) to keep the timings consistent.

```
require(expm)

## Loading required package: expm
## Loading required package: Matrix
##
## Attaching package: 'expm'
##
## The following object is masked from 'package:Matrix':
##
## expm

tt <- c(0.77, 1.69, 2.69, 3.67, 4.69, 5.71, 7.94, 9.67, 11.77, 17.77, 23.77,
        32.77, 40.73, 47.75, 54.9, 62.81, 72.88, 98.77, 125.92, 160.19, 191.15,
        223.78, 287.7, 340.01, 340.95, 342.01)
y <- c(1.396, 3.784, 5.948, 7.717, 9.077, 10.1, 11.263, 11.856, 12.251,
       12.699, 12.869, 13.048, 13.222, 13.347, 13.507, 13.628, 13.804, 14.087,
       14.185, 14.351, 14.458, 14.756, 15.262, 15.703, 15.703, 15.703)
ones <- rep(1, length(t))

Mpred <- function(theta) {
    # WARNING: assumes tt global
    kvec <- exp(theta[1:3])
    k1 <- kvec[1]
    k2 <- kvec[2]
    k3 <- kvec[3]
    # MIN problem terbuthylazene disappearance
    z <- k1 + k2 + k3
    y <- z * z - 4 * k1 * k3
    l1 <- 0.5 * (-z + sqrt(y))
    l2 <- 0.5 * (-z - sqrt(y))
    val <- 100 * (1 - ((k1 + k2 + l2) * exp(l2 * tt) - (k1 + k2 + l1)
            * exp(l1 * tt))/(l2 - l1))
} # val should be a vector if t is a vector

negll <- function(theta) {
    # non expm version JN 110731
    pred <- Mpred(theta)
    sigma <- exp(theta[4])
    -sum(dnorm(y, mean = pred, sd = sigma, log = TRUE))
}

nlogL <- function(theta) {
    k <- exp(theta[1:3])
    sigma <- exp(theta[4])
    A <- rbind(c(-k[1], k[2]), c(k[1], -(k[2] + k[3])))
    x0 <- c(0, 100)
```

```
    sol <- function(tt) {
        100 - sum(expm(A * tt) %*% x0)
    }
    pred <- sapply(tt, sol)
    -sum(dnorm(y, mean = pred, sd = sigma, log = TRUE))
}
mytheta <- c(-2, -2, -2, -2)
vnlogL <- nlogL(mytheta)
cat("nlogL(-2,-2,-2,-2) = ", vnlogL, "\n")

## nlogL(-2,-2,-2,-2) =  3292470

vnegll <- negll(mytheta)
cat("negll(-2,-2,-2,-2) = ", vnegll, "\n")

## negll(-2,-2,-2,-2) =  3292470
```

We now have two ways to compute the negative log likelihood, nlogL() and negll(). We have, furthermore, checked that they give the same answer for a set of parameters. The nlogL() function uses the expm() exponential matrix function, which is notationally very convenient but notoriously difficult to compute reliably (Moler and Loan, 2003). There are two sets of tools in R to compute it, one in the **Matrix** package and the other in the package **expm** that has several approaches (we will use the default chosen by each package). tve

```
##                time.nlogL time.negll    ratio
## regular          55612122    25497.7 2181.067
## compiled         55412567    25679.3 2157.869
## regular +expm    14615661    26292.7  555.884
## compiled+expm    14571569    26148.5  557.262
```

While it is clear that the **expm** code is much more efficient than that in **Matrix**, avoiding the matrix exponential is a very very good idea if we want an efficient computation. Moreover, in this case, we see very little benefit from the compiler.

18.3.4   Specializing codes

We have mentioned in Chapter 11 that bounds constraints are very useful for the reliability and usefulness of function minimization programs. However, we should note that the bounds constraint feature does not come for free. In my packages **Rvmmin** and **Rcgmin**, I check for bounds and then call separate functions for the bounded or unconstrained cases. Let us check on a problem that is unconstrained whether this is worthwhile.

```
require(microbenchmark)
require(Rvmmin)
require(Rcgmin)
sqs <- function(x) {
    sum(seq_along(x) * (x - 0.5 * seq_along(x))^2)
}
```

```
sqsg <- function(x) {
    ii <- seq_along(x)
    g <- 2 * ii * (x - 0.5 * ii)
}
xstrt <- rep(pi, 20)
lb <- rep(-100, 20)
ub <- rep(100, 20) # very loose bounds
bdmsk <- rep(1, 20) # free parameters
tvmu <- microbenchmark(avmu <- Rvmminu(xstrt, sqs, sqsg))
tcgu <- microbenchmark(acgu <- Rcgminu(xstrt, sqs, sqsg))
tvmu <- microbenchmark(avmb <- Rvmminb(xstrt, sqs, sqsg, bdmsk = bdmsk,
lower = lb, upper = ub))
tcgb <- microbenchmark(acgb <- Rcgminb(xstrt, sqs, sqsg, bdmsk = bdmsk,
lower = lb, upper = ub))
svmu <- summary(tvmu$time)
cnames <- names(svmu)
svmu <- as.numeric(svmu)
svmb <- as.numeric(summary(tvmb$time))
scgu <- as.numeric(summary(tcgu$time))
scgb <- as.numeric(summary(tcgb$time))
names(svmu) <- cnames
names(svmb) <- cnames
names(scgu) <- cnames
names(scgb) <- cnames
mytab <- rbind(svmu, svmb, scgu, scgb)
rownames(mytab) <- c("Rvmminu", "Rvmminb", "Rcgminu", "Rcgminb")
# dput(mytab, file='includes/C18mytabJ6.dput')
ftable(mytab)
```

```
##              Min.  1st Qu.   Median      Mean  3rd Qu.      Max.
##
## Rvmminu   5621000  5738000  5844000   6290000  7060000   8447000
## Rvmminb  26590000 28160000 28470000  28680000 28930000  49850000
## Rcgminu   2466000  2517000  2557000   2723000  2604000   6137000
## Rcgminb  23270000 24020000 24790000  24780000 25180000  30120000
```

We see that the bounds constrained versions of the codes are four to nine times slower on this simple function, where we can anticipate that the bounds were never active.

18.4   External language compiled functions

We can also consider interfacing our optimizers to objective functions compiled in other computing languages. While interfacing R to a Fortran, C, or other compiled-code optimizer is relatively tedious and involved, using a compiled objective function in such languages is quite easy. To illustrate this, let us use the minimization of the Rayleigh quotient to find the largest eigenvalue of a symmetric matrix. The example we will use is the Moler matrix from Appendix 1 of Nash (1979), which is defined in the following code.

```
molermat <- function(n) {
    A <- matrix(NA, nrow = n, ncol = n)
    for (i in 1:n) {
        for (j in 1:n) {
            if (i == j)
                A[i, i] <- i else A[i, j] <- min(i, j) - 2
        }
    }
    A
}

rayq <- function(x, A) {
    rayquo <- as.numeric(crossprod(x, crossprod(A, x))/crossprod(x))
}
```

We will also compute the eigensolutions of the Moler matrix using R's function eigen(), which finds all eigenvalues and eigenvectors.

```
require(microbenchmark)
require(compiler)
molermatc <- cmpfun(molermat)
rayqc <- cmpfun(rayq)
n <- 200
x <- rep(1, n)
tbuild <- microbenchmark(A <- molermat(n))$time
tbuildc <- microbenchmark(A <- molermatc(n))$time
teigen <- microbenchmark(evs <- eigen(A))$time
tr <- microbenchmark(rayq(x, A))$time
trc <- microbenchmark(rayqc(x, A))$time
eigvec <- evs$vectors[, 1] # select first eigenvector
# normalize first eigenvector
eigvec <- sign(eigvec[[1]]) * eigvec/sqrt(as.numeric(crossprod(eigvec)))
```

The timings are summarized below in Section 18.4.2, showing that building the matrix takes a considerable computational effort and can be made quicker with the byte compiler. However, the speedup in the evaluation of the Rayleigh quotient with the byte compiler is negligible.

Once we have matrix A, we can try finding the maximum eigensolution by minimizing the Rayleigh quotient for matrix -A. To make the eigenvectors comparable between solution methods, we need to normalize them. I have chosen to scale the eigenvector so the sum of squares of the its components is 1 and the first component is positive. We compare the eigenvalue and vector with the comparable result from eigen().

Anticipating the timings we report later based on the time we must wait for the optimization to run, it is clear that this approach to the eigensolution is not competitive in timing with eigen().

```
require(optimx)
tcg <- system.time(amax <- optimx(x, fn = rayq, method = "Rcgmin", control =
list(usenumDeriv = TRUE,
    kkt = FALSE), A = -A))[[3]]
# summary(amax)
```

```
cat("maximal eigensolution: Value=", -amax$value, "  timecg= ", tcg, "\n")
tvec <- coef(amax)[1, ]
rcgvec <- sign(tvec)[[1]] * (tvec)/sqrt(as.numeric(crossprod(tvec)))
cat("Compare with eigen()\n")
cat("Difference in eigenvalues = ", (-amax$value[1] - evs$values[[1]]), "\n")
cat("Max abs vector difference = ", max(abs(eigvec - rcgvec)), "\n")
```

We could use the byte compiler on the function `rayq()`, but I found very little advantage in so doing. Possibly as *n* gets larger, the advantage of the compiled code could be more pronounced. This is using an explicit matrix A. Given that we know the matrix elements, we could implicitly perform the matrix multiplication. Here is the code to do that.

```
axmoler <- function(x) {
    # A memory-saving version of A%*%x For Moler matrix.
    n <- length(x)
    j <- 1:n
    ax <- rep(0, n)
    for (i in 1:n) {
        term <- x * (pmin(i, j) - 2)
        ax[i] <- sum(term[-i])
    }
    ax <- ax + j * x
    ax
}
require(compiler)
require(microbenchmark)
n <- 200
x <- rep(1, n)
tax <- microbenchmark(ax <- axmoler(x))$time
axmolerc <- cmpfun(axmoler)
taxc <- microbenchmark(axc <- axmolerc(x))$time
cat("time.ax, time.axc:", mean(tax), mean(taxc), "\n")
## On J6 gives time.ax, time.axc: 5591866 4925221
```

Again, the byte compiler is not making a lot of difference, although we have avoided storing a large matrix. Of itself, using an implicit multiplication in R may not be helpful for speeding up computations. The issue is that we have to essentially create the matrix each time we do a matrix–vector product. However, if we embed such a computation in a faster computing environment, we do get some gains. Let us look at running the matrix–vector and the Rayleigh quotient calculations in Fortran.

18.4.1   Building an R function using Fortran

While we show Fortran here, the same general approach will hold for C or other compatible compiled languages.

First, we need to write a Fortran **subroutine** because R can only call routines with a void return. Here is an example subroutine that multiplies a vector `x` by the Moler matrix implicitly giving the result in vector `ax`. It then computes the Rayleigh quotient, returning the result in the argument variable `rq`.

```
c      This is file rqmoler.f
       subroutine rqmoler(n, x, rq)
       integer n, i, j
       double precision x(n), ax(n), sum, rq
c      return rq = t(x) * A * x / (t(x) * x)  for A = moler matrix
c      A[i,j]=min(i,j)-2 for i<>j, or i for i==j
       do 20 i=1,n
          sum=0.0
          do 10 j=1,n
             if (i.eq.j) then
                sum = sum+i*x(i)
             else
                sum = sum+(min(i,j)-2)*x(j)
             endif
 10       continue
          ax(i)=sum
 20    continue
       rq=0.0
       sum=0.0
       do 30 i=1,n
          rq = rq+ax(i)*x(i)
          sum = sum + x(i)**2
 30    continue
       rq = rq/sum
       return
       end
```

Second, we need to build the dynamic load library. R provides a tool for this that is run from the command line as follows.

```
R CMD SHLIB rqmoler.f
```

This will create files rqmoler.o and rqmoler.so in the working directory for Linux/Unix systems. On Microsoft Windows machines (I have only tested this in a Windows XP virtual machine), the user must open a command window and issue the same command to create a dynamic link library file rqmoler.dll. As far as I can determine, the **Rtools** collection by Duncan Murdoch is sufficient for this task.

We now need to load this routine into our R session, which is done with

```
dyn.load("rqmoler.so")
## In Windows dyn.load("rqmoler.dll")
cat("Is the Fortran RQ function loaded ",is.loaded('rqmoler'),"\n")
```

and we check that the Fortran routine is available using the is.loaded() function. Note that we leave off the filename suffix.

The function "rqmoler" takes a vector x and returns the Rayleigh quotient. But it is important to note that we do not need the matrix A at all. Note that the Fortran rqmoler.f returns a list of elements n, x, and rq, where the last item is the result we want.

```
dyn.load("/home/john/nlpor/supportdocs/tuning/rqmoler.so")
cat("Is the Fortran RQ function loaded? ", is.loaded("rqmoler"), "\n")

## Is the Fortran RQ function loaded?  TRUE
```

We need to get the negative of the Rayleigh quotient for our maximal Eigensolution and then optimize it, which we do with the following code.

```
frqmoler <- function(x) {
    n <- length(x)
    rq <- 0
    res <- .Fortran("rqmoler", n = as.integer(n), x = as.double(x),
rq = as.double(rq))
    res$rq
}
nfrqmoler <- function(x) {
    (-1) * frqmoler(x)
}
tcgrf <- system.time(amaxi <- optimx(x, fn = nfrqmoler, method = "Rcgmin",
control = list(usenumDeriv = TRUE,
    kkt = FALSE, trace = 1)))[[3]]
```

18.4.2   Summary of Rayleigh quotient timings

Noting that these examples are, of course, just examples, let us look at the timings for different tasks in the Rayleigh quotient minimization for the Moler matrix of dimension 200 on our machine named J6

- In R, just building the Moler matrix of order 200 takes on average 147.857 ms, but the byte compiler reduces this dramatically to 42.787 ms.

- Our main comparison is with the function eigen(), which computes all eigensolutions in a mean time of 30.614 ms. This is very fast and hard to beat. Note that the overall time should add in the "build" if we compare with methods that develop the matrix implicitly.

- Once we have the matrix A, then computing the Rayleigh quotient in R takes an average of 0.215 ms, interpreted, or 0.201 ms, byte compiled.

- If we use the Moler matrix implicitly to compute the Rayleigh quotient in R, this takes an average of 5.602 ms, interpreted, or 5.291 ms, byte compiled.

- Using a Fortran version of the Rayleigh quotient called through frqmoler(), where the matrix is generated implicitly takes about 0.083 ms. For compatibility with a calling program that specifies the matrix, I actually tested a version of this with the matrix in the argument list but found almost identical timings. That is, I specified

```
frqmolerA<-function(x, A)
```

in a definition otherwise identical to function `frqmoler()`. The matrix is, of course, not used. The Fortran function is over 60 times faster than the R one using an implicit matrix. Ignoring the building time, it is still better than double the time using an explicit matrix.

- To minimize the Rayleigh quotient for the maximal eigenvalue takes quite a long time, indeed over 4 s (not milliseconds) using the explicit matrix form and, worse, nearly 50 s with the implicit Rayleigh quotient. We do better with the Fortran implicit version but still take over 1 s. And this is to get just a single eigensolution, where `eigen()` gets them all in about 30 ms.

From the above, it may appear that minimizing the Rayleigh quotient is unsuitable for computing eigensolutions. In general I would agree, but there are specialized minimization tools (Geradin, 1971) that, coded in Fortran or some other fast computational environment, do very well. However, such a method is unlikely to be a choice of R users.

It is worth noting that each situation that must be speeded up requires at least some individual attention. From a more extensive set of timings, now over a year old, I will venture the following generalizations:

- A matrix–vector multiplication function that implicitly performs the operation can be written so that no actual matrix is needed. This saves a great deal of memory when the order of the matrix gets large, but for the Moler matrix uses element operations. This gains from compilation either with `cmpfun` or with an external language such as Fortran. However, the R version of implicit matrix–vector routine is much slower than the Fortran one.

- Compilation with `cmpfun()` provides quite decent time reductions for building the matrix (`molermat`). That is, compilation helps reduce operations on elements of vectors and matrices. However, these operations remain relatively slow in R.

- Compiling an optimizer like `spg` from package **BB** that already is vector oriented and gave almost negligible speedup.

Both Ravi Varadhan and Gabor Grothendieck provided valuable input to the work that underlies the discussion of the Rayleigh quotient problem.

18.5   Deciding when we are finished

One of the major parts of any iterative algorithm is deciding when it is finished, or rather when the approximation to the answer we have is "good enough." In Chapter 17, we considered various conditions that good optimization solutions should satisfy. These are mainly that the gradient or some measure thereof is "small," while the surface around the solution curves upwards. Actually computing such tests, as we saw, can be computationally demanding, so within optimization codes we often look for proxies for the actual gradient and particularly for Hessian information.

As mentioned in Chapter 2, gradient methods generally stop when the norm of the gradient is small. While some developers are happy to use the gradient projection along the current search direction for this measure, I like to revert to the steepest descents direction and stop only when no progress can be made with that as our search vector. However, this does sometimes cause extra work to be performed that is not necessary.

Note that bounds or other constraints render the termination tests more complicated, as we must first project our search onto the constraint surface. For bounds, the changes are relatively simple but do need care. Surprisingly, masks, for which the changes to code are almost trivial, seem to be rarely offered in optimizer codes. In R we need to recognize that both bounds and masks involve individual matrix or vector elements and may degrade performance compared to fully vectorized codes.

One of the major decisions that developers must make is "how small is zero" in a particular problem. Choosing tolerances too small will make an algorithm try very hard to reduce the objective function and use unnecessary computational effort. But tolerances that are too large risk returning "solutions" that are far from a true optimum. My own preference is to err on the side of caution and make the tolerances rather strict. Users who really understand their problem can relax the tolerances.

Mostly I prefer to avoid tolerances, as users are prone to change them and then complain to me that "your program doesn't work" because it has stopped well before it has finished and stopped because they changed the tolerance. Therefore, I often will test for "no change" in numbers by testing for equality between a + offset and b + offset where offset is a modest number of the order of 100 to 10 000 for regular `double` numbers. When a and b are small, the numbers compared eventually become just the value of offset, while when a and b are quite large, we are comparing their most significant digits. Users generally do not tinker with the offset, as it does not appear to be a form of tolerance.

If the methods are entirely coded in R, users can modify the actual R code of the termination tests if they must solve many similar problems. I advise this to be investigated if there could be benefits in a production-type environment. Additionally, the entire objective and optimizer may be implemented in a compiled language and called from R as above. R can still provide convenience in loading and transforming data and in drawing graphs of data or results.

For exploratory and interactive work, I generally advise very tight tolerances. The reason for this is to force optimizers to strive for the best answer possible, but it comes at the price of some, possibly many, extra iterations and function and gradient evaluations. My own codes tend to rather aggressive tolerances and I believe that they mostly do much more work than they need. On the other hand, if they can continue to reduce the objective in approaching a solution, they do.

18.5.1   Tests for things gone wrong

Sometimes a calculation is "finished" because the user has supplied some input that is inappropriate for the situation. "NA" values (missing data) may be acceptable in some situations but is rarely appropriate for parameters of objective functions.

Similarly, if the initial function value computation is attempted at an inadmissible point, for example, where there is a zero divide or square root of a negative number, then there is no sense in proceeding. However, within the optimization, as we have noted in Chapter 3, trial sets of parameters that are inadmissible may be generated. In some methods, we may be able to recognize this situation and continue with the optimization. However, when progress is not possible, we need to terminate with notification to the user. See section 3.7.

`optplus` included a number of checks that our objective function returns suitable results. We want a numeric scalar result. Some optimizers will still "work" with other objective function types, but the ultimate answers could be unreliable. Unfortunately, carrying out all these tests takes time. In fact, it takes so much time that I have set aside that direction of development for my optimization tools. Moreover, if you can ensure the inputs to functions are appropriate, then it is worth looking into how you can bypass such checks.

It is worth noting that a particular case that has caused trouble over the years is the occurrence of objects that are effectively 1×1 matrices. These behave "mostly" like scalar numbers in computations, but there are enough functions that cannot deal with them to get us into trouble. If in doubt, convert these to numbers with the `as.numeric()` function. More generally, many functions that deal with matrices or vectors are not set up to deal with the $n = 1$ situation.

From the above, you may correctly surmise that a modest level of paranoia is useful when writing our objective functions and code to launch the optimizations, as errors creep in very easily.

References

Dongarra JJ, Luszczek P, and Petitet A 2003 The LINPACK benchmark: past, present, and future. Concurrency and Computation: Practice and Experience 15(9), 803–820.

Eugster MJA and Leisch F 2011 Exploratory analysis of benchmark experiments – an interactive approach. Computational Statistics 26(4), 699–710.

Eugster MJA, Hothorn T, and Leisch F 2012 Domain-based benchmark experiments: exploratory and inferential analysis. Austrian Journal of Statistics 41(1), 5–26.

Geradin M 1971 The computational efficiency of a new minimization algorithm for eigenvalue analysis. Journal of Sound and Vibration 19, 319–331.

Kusnierczyk W 2012 *rbenchmark: Benchmarking routine for R*. R package version 1.0.0.

Mersmann O 2013 *microbenchmark: sub microsecond accurate timing functions*. R package version 1.3-0.

Moler C and Loan CV 2003 Nineteen dubious ways to compute the exponential of a matrix, twenty-five years later. SIAM Review 45(1), 3–49.

Nash JC 1979 Compact Numerical Methods for Computers: Linear Algebra and Function Minimisation. Adam Hilger, Bristol. Second Edition, 1990, Bristol: Institute of Physics Publications.

19

Linking R to external optimization tools

This chapter is about using tools that are **NOT** in R but prepared in other computing systems. However, it is about using R to access these tools.

The issues in doing this are first of all technical issues. We must find a way to link our R session to the external tool in a way that the tool can be run successfully and we can get the results back in a form usable by R. There are several ways to do this, and the level of detail is such that here we will have to be satisfied with a limited treatment of the main ideas.

There are also legal issues that may concern us. R is free/libre software, and proprietary or otherwise limited licenses on the external tool mean that such tools cannot be distributed with R. Such issues occupy quite a large number of the contributions on the R mailing lists and to some extent fragment the R environment. Personally, I will only work with free/libre software unless there is some compelling reason to do otherwise. I am a scientist and want to share ideas with others, and this cannot be done properly unless source code and data are available so that computations can be reproduced.

Finally, there are personal preferences. My own choices are, as far as is reasonable, to avoid complexity that can give rise to errors. In any situation where one has to deal with more than one structure or system, such errors can be extremely difficult to debug.

Nevertheless, there are situations where it is necessary to use external tools, and this chapter discusses some of the approaches that can be taken.

Nonlinear Parameter Optimization Using R Tools, First Edition. John C. Nash.
© 2014 John Wiley & Sons, Ltd. Published 2014 by John Wiley & Sons, Ltd.
Companion Website: www.wiley.com/go/nonlinear_parameter

19.1   Mechanisms to link R to external software

To provide a convenient way to discuss interfacing to external tools, let us suppose we wish to use either an external program called xprog or an external subroutine xsub or a function xfun. Clearly, we should be able to fairly easily prepare an xprog if we have either of the other two forms.

19.1.1   R functions to call external (sub)programs

R has a number of functions (see R Core Team, 2013) to invoke functions or subroutines in other programming languages.

- .Fortran
- .C
- .Call
- .External
- dyn.load

I have never used any of these other than .Fortran and dyn.load as presented in Section 18.4 to compute an objective function. As described in the example in Section 18.4.1, this can be highly effective in increasing performance of optimization methods, because the evaluation of the objective or other functions is generally a very large component of the overall execution time.

19.1.2   File and system call methods

We can avoid many of the debugging tasks between R and external tools by having R prepare an input file for the external program xprog to read and process. The big advantage of this is that we can separately verify both sides of the transaction. The R system() command can be used to invoke xprog, so we do not ultimately have to manually invoke the program. Moreover, xprog can write its results to a file that R can read and use.

Obviously, this approach involves potentially slow disk write and read operations, and there is a level of tedium to prepare code to appropriately format data that must be written and equally to read the results from xprog. Nevertheless, this approach is the one I generally consider first because it is the most reliable to implement. Moreover, it allows the work to be shared among several workers, because there is a natural separation of the tasks. In some cases, performance can be improved to a satisfactory level by using a simulated or RAM disk or memory disk to streamline the "disk" operations.

19.1.3   Thin client methods

Sometimes there are services we can use to do our computations. For optimization, the best known of these is the NEOS server (http://www.neos-server.org/). In many respects, these approaches are similar to those of the previous section, because we must prepare data to send to the server and then process the returned information.

19.2   Prepackaged links to external optimization tools

19.2.1   NEOS

A package to use the NEOS server mentioned earlier has been prepared for R but unfortunately does not appear to have a vignette, demo() or end-to-end example, which is a pity. Like many (most?) workers in computation, I tend to learn and use a piece of software by modifying a working example. The raw package, even with its documentation, seems less user friendly than the Web interface provided by the NEOS operators.

19.2.2   Automatic Differentiation Model Builder (ADMB)

Automatic Differentiation Model Builder (ADMB, see admb-project.org) is closer to the kind of modeling done with R via tools like nls() or some of the maximum likelihood packages. We have already mentioned it in Chapters 3 and 12. For some types of models, ADMB offers significant advantages over other tools, including R. It turns out that there are two R packages, both on CRAN, that interface with ADMB. These are **PBSadmb** (Schnute et al. 2013) and **R2admb** (Bolker et al. 2013). It happens that I have had occasion to exchange ideas with both Jon Schnute and Ben Bolker for many years and have a great deal of respect for both.

R2admb has a vignette, which is extremely useful starting point for working with ADMB. On the other hand, **PBSadmb** offers a GUI to help build solutions. Unfortunately, the installation of either package in R does not check that ADMB software is itself installed and configured, and I found that to be a task that was unexpectedly different from my expectations. I was able to guess how to get things working based on my experience with Linux software over the years, but I suspect many users would be frustrated. As the software does, in fact, work and work well, this is more a warning about the difficulties that can arise in using external software rather than a complaint about ADMB itself. It really just needs a bit more documentation or slightly fancier packaging. The documentation may also work just fine for Windows users, although they will have to also install an appropriate compiler, because ADMB creates **C++** code that is compiled and run to perform the estimation tasks.

Both packages unfortunately require a file of type .tpl to define a problem for ADMB. ADMB then processes this file. Users may wish for software that allows a point and click mechanism to define such a file. That, of course, involves a non-trivial set of tasks, but it is the genre of software that nonspecialist users seek.

19.2.3   NLopt

NLopt (Johnson, 2008) is a fairly comprehensive set of optimization tools, some of which already have counterparts or implementations in R. We have already mentioned this set of tools in Chapter 9. There are two closely related R packages for using **NLopt** in R:

- **nloptr** (Ypma, 2013) is a more or less straightforward interface to the underlying **NLopt** capabilities, while

- **nloptwrap** (Borchers, 2013) changes the syntax of the calls in **nloptr** to a form that resembles that familiar to users of optim() and other optimizers.

Although these packages are interfaces to external tools, a nice feature is that installing them also installs the underlying external programs. That is, we do not need to separately install the external tool.

19.2.4   BUGS and related tools

Baysian inference Using Gibbs Sampling (BUGS) (Lunn et al. 2009) is a Markow Chain Monte Carlo method of stochastically estimating certain types of models, which may include nonlinear models. This is a very different approach to that used elsewhere in this book, and I do not propose to present details. That such an approach is of interest to many workers is illustrated, however, by the number of R packages to interface to the various OpenBUGS, WinBUGS, or JAGS (Just Another Gibbs Sampler) tools. I have only used JAGS, and only for a single example problem, so am not qualified to speak to the merits of this approach.

To use these tools, however, it is necessary to first install the relevant software, which requires some care and knowledge. Furthermore, while R can call the BUGS/JAGS software, it works through a model that is expressed in the form of files particular to that software.

19.3   Strategy for using external tools

My own strategy for using external tools is

- If possible, use tools that are properly embedded in R or at least are fully integrated with the install.packages() infrastructure so that separate installation of the external tool is avoided.

- Where it is necessary to use an external tool, it is worth structuring the interface so that R and the external tool are run more or less independently and data is passed through files.

- As demonstrated in Chapter 18, writing the objective function in a compiled programming language like Fortran can give a remarkable performance gain.

References

Bolker B, Skaug H, and Laake J 2013 R2admb: ADMB to R interface functions. R package version 0.7.10.

Borchers HW 2013 nloptwrap: Wrapper for Package nloptr. R package version 0.5-1.

R Core Team 2013 Writing R Extensions.

Johnson SG 2008 The NLopt nonlinear-optimization package.

Lunn D, Spiegelhalter D, Thomas A, and Best N 2009 The bugs project: evolution, critique and future directions. Statistics in Medicine 28(25), 3049–3067.

Schnute JT, Haigh R, and Couture-Beil A 2013 PBSadmb: ADMB for R Using Scripts or GUI. R package version 0.66.87.

Ypma J 2013 nloptr: R interface to NLopt. R package version 0.9.3.

20

Differential equation models

This chapter follows steps that attempt to illustrate the estimation of an SIWR model from the problem proposed by Marisa Eisenberg in "Lab Session: Intro to Parameter Estimation & Identifiability, 6−17−13." This serves as a useful overview of dealing with differential equation models and their estimation.

20.1   The model

The description here refers directly to the program code below. The model aims to explain the numbers of susceptible (S), infected (I), and recovered (R) in a cholera outbreak. The particular model includes a variable that measures bacteria concentration in water (W). This leads to the SIWR model developed by Tien and Earn (2010), Bi and Bw are parameters that describe the direct (human−human) and indirect (waterborne) cholera transmission. e is the pathogen decay rate in the water.

We do not measure I directly, but a scaled version of it

```
y = I/k
```

where k is a combination of the reporting rate, the asymptomatic rate, and the total population size.

Our equations include a recovery rate and a birth−death parameter, but in the present situation, the description of the problem states that the former is fixed at 0.25 and the latter is 0.

Unfortunately, the code suggested fails to do a good job of optimizing the cost function. An e-mail exchange between myself and Nathan Lemoine that prompted this investigation was initiated with his blog posting http://www.r-bloggers.com/evaluating-optimization-algorithms-in-matlab-python-and-r/.

Nonlinear Parameter Optimization Using R Tools, First Edition. John C. Nash.
© 2014 John Wiley & Sons, Ltd. Published 2014 by John Wiley & Sons, Ltd.
Companion Website: www.wiley.com/go/nonlinear_parameter

20.2   Background

Ecological models that involve multiple differential equations are common. They are closely related to compartment models for pharmacokinetics and chemical engineering.

The motivating problem involves a series of measurements for different time points. Unfortunately, after this section was written, we realized that the data set used had not been formally released for use, so I have replaced it with a similar set of data that has been generated by loosely drawing a similar picture and "guesstimating" the numbers from that graph.

The variable data is observed for time tspan. We load the package **deSolve** and establish the ordinary differential equations (ODEs) in the function SIRode. Some housekeeping matters include merging the model parameters into a vector (params) and specifying initial conditions for the differential equations that depend on one of these parameters (k). Noting that the initial guess to parameter k is very small, it is sensible to work with a scaled parameter, and our choice is to use 1e+5*k as the quantity to estimate. Moreover, e is initially set at 0.01, so let us use 100*e as the quantity to optimize.

The suitability and motivations for the present model will not be argued here. These are important matters, but they are somewhat separate from the estimation task.

```
tspan <- c(0, 8, 13, 20, 29, 34, 41, 50, 56, 62, 71, 78, 83, 91, 99, 104, 111,
    120, 127, 134, 139, 148, 155, 160)
data <- c(115, 59, 69, 142, 390, 2990, 4700, 5488, 5620, 6399, 5388, 4422,
    3566, 2222, 1911, 2111, 1733, 1166, 831, 740, 780, 500, 540, 390)
## Main code for the SIR model example
require(deSolve) # Load the library to solve the ode

## Loading required package: deSolve

## initial parameter values ##
Bi <- 0.75
Bw <- 0.75
e <- 0.01
k <- 1/89193.18
## define a weight vector OUTSIDE LLode function
wts <- sqrt(sqrt(data))
## Combine parameters into a vector
params <- c(Bi, Bw, e * 100, k * 1e+05)
names(params) <- c("Bi", "Bw", "escaled", "kscaled")
## ODE function ##
SIRode <- function(t, x, params) {
    S <- x[1]
    I <- x[2]
    W <- x[3]
    R <- x[4]
    Bi <- params[1]
    Bw <- params[2]
    e <- params[3] * 0.01
    k <- params[4] * 1e-05
```

```
    dS <- -Bi * S * I - Bw * S * W
    dI <- Bi * S * I + Bw * S * W - 0.25 * I
    dW <- e * (I - W)
    dR <- 0.25 * I
    output <- c(dS, dI, dW, dR)
    list(output)
}
## initial conditions ##
I0 <- data[1] * k
R0 <- 0
S0 <- (1 - I0)
W0 <- 0
initCond <- c(S0, I0, W0, R0)
```

Let us use the initial parameter guesses above and solve the ODEs using a particular method (Runge–Kutta(4,5) method) which is suitable for a fairly wide range of differential equations. We will graph the data and the model that use the parameter guesses (Figure 20.1), where the dashed line is the model.

```
## Run ODE using initial parameter guesses ##
initSim <- ode(initCond, tspan, SIRode, params, method = "ode45")
plot(tspan, initSim[, 3]/k, type = "l", lty = "dashed")
points(tspan, data)
```

Clearly, the model does not come close enough to the data to be useful. We have some improvements to make in the parameter values.

There is also the possibility that we could use a different integrator for the differential equations. What is the justification for choosing the Runge–Kutta(4, 5) method? Admittedly, this is a popular and reliable method, and its wide use attests to its general strength. However, there is also the LSODA solver that is intended to balance between stiff and nonstiff solvers, as well as a number of other choices. In this area, I count myself essentially a novice. In the following function LLode, ode45 is used, as suggested in the original script sent to me.

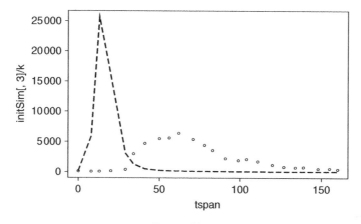

Figure 20.1

20.3   The likelihood function

The suggested solution involves maximizing the likelihood or, for our purposes, minimizing the negative log likelihood. Let us first define this. In the present case, it is no more than a weighted sum of squares. However, to get the model, we need to solve the differential equations repeatedly. Moreover, the original code discussed in the blog that motivated this article did not check the diagnostic flags of the solver. We now do that. Note that the value of the flag may change according to the solver selected within the **deSolve** package.

It is arguable that the decision to allow different "success" code values is inviting users to get into trouble. The reason is probably that the codes assembled together and wrapped into **deSolve** return these codes, and it is useful for the developer to compare results from the wrapper with those from the original code in its original programming language.

```
## Define likelihood function using weighted least squares ##
LLode <- function(params) {
    k <- params[4] * 1e-05
    I0 <- data[1] * k
    R0 <- 0
    S0 <- 1 - I0
    W0 <- 0
    initCond <- c(S0, I0, W0, R0)
    # Run the ODE
    odeOut <- ode(initCond, tspan, SIRode, params, method = "ode45")
    if (attr(odeOut, "istate")[1] != 0) {
        ## Check if satisfactory 2 indicates perceived success of
        method 'lsoda', 0 for
        ## 'ode45'. Other integrators may have different indicator codes
        cat("Integration possibly failed\n")
        LL <- .Machine$double.xmax * 1e-06 # indicate failure
    } else {
        y <- odeOut[, 3]/k # Measurement variable
        wtDiff <- (y - data) * wts # Weighted difference
        LL <- as.numeric(crossprod(wtDiff)) # Sum of squares
    }
    LL
}
```

20.4   A first try at minimization

Let us try to minimize this function using R's optim(), which will invoke the Nelder–Mead method by default, although we make the call explicit in the following. We could use trace=1 to observe the minimization.

```
## optimize using optim() ##
MLoptres <- optim(params, LLode, method = "Nelder-Mead",
    control = list(trace = 0, maxit = 5000))
MLoptres
```

```
## $par
##        Bi        Bw   escaled   kscaled
## 0.307331 1.142383 0.453779 0.965853
##
## $value
## [1] 208064645
##
## $counts
## function gradient
##      221       NA
##
## $convergence
## [1] 0
##
## $message
## NULL
```

We note a considerably better fit to the data with the optimized parameters, and this is reflected in Figure 20.2 (solid line), where we include the model using the preliminary guesses to the parameters (dotted line).

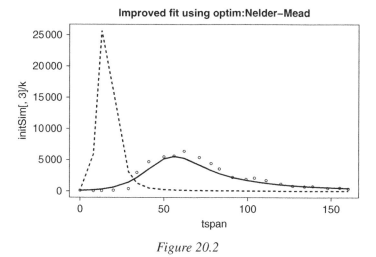

Figure 20.2

20.5   Attempts with `optimx`

Let us try some methods in the package **optimx**. We now add some bounds to the parameters and choose two methods that are suitable for bounds and do not need derivatives. After this, we see that there is a very small – almost invisible – improvement in the model. As this computation is quite slow – look at the timings displayed with the optimizers in column `xtimes` – we ran this off-line and saved the result using the R command `dput()`.

```
require(optimx)
optxres <- optimx(params, LLode, lower = rep(0, 4), upper = rep(500, 4),
method = c("nmkb", "bobyqa"), control = list(usenumDeriv = TRUE,
maxit = 5000, trace = 0))
summary(optxres, order = value)
## dput(optxres, file='includes/C20opresult.dput')


bestpar <- coef(summary(optxres, order = value)[1, ])
cat("best parameters:")
print(bestpar)
## dput(bestpar, file='includes/C20bestpar.dput')


bpSim <- ode(initCond, tspan, SIRode, bestpar, method = "ode45")
X11()
plot(tspan, initSim[, 3]/k, type = "l", lty = "dashed")
points(tspan, data)
points(tspan, MLSim[, 3]/k, type = "l", lty = "twodash")
points(tspan, bpSim[, 3]/k, type = "l")
title(main = "Improved fit using optimx")



## from running supportdocs/ODEprobs/ODElemoine.R
optxres <- dget("includes/C20opresult.dput")
optxres


##              Bi        Bw escaled     kscaled       value fevals gevals niter
## bobyqa 0.3060053   1.87848 0.283434  0.981722   207751806   5000     NA    NA
## nmkb   0.0352253 493.81805 3.061697 114.234193 14907264560    585     NA    NA
##        convcode  kkt1  kkt2   xtimes
## bobyqa        0 FALSE FALSE   29.010
## nmkb          0 FALSE FALSE  104.606


bestpar <- dget("includes/C20bestpar.dput")
```

20.6   Using nonlinear least squares

Given the form of the likelihood, we could use a nonlinear least squares method.
Here is the residual function.

```
Lemres <- function(params) {
    k <- params[4] * 1e-05
    I0 <- data[1] * k
    R0 <- 0
    S0 <- 1 - I0
    W0 <- 0
    initCond <- c(S0, I0, W0, R0)
    nobs <- length(tspan)
    resl <- rep(NA, nobs)
    # Run the ODE
    odeOut <- ode(initCond, tspan, SIRode, params, method = "ode45")
    if (attr(odeOut, "istate")[1] != 0) {
        ## Check if satisfactory
        cat("Integration possibly failed\n")
        resl <- rep(.Machine$double.xmax * 1e-12, nobs)
    } else {
```

```
        y <- odeOut[, 3]/k
        resl <- (y - data) * wts
    }
    resl
}
```

Now we will run the nonlinear least squares minimizer.

```
## Loading required package: nlmrt

## nlmrt class object: x
## residual sumsquares =  218712359  on  24 observations
##      after  15    Jacobian and  24 function evaluations
##    name       coeff        SE       tstat       pval     gradient    JSingval
## Bi        0.330594    0.01157      28.59           0    2.134e+09     2102404
## Bw        0.146237    0.07912       1.848     0.07939     80497932      234087
## escaled    1.26505      2.152      0.5878      0.5632     12993872       44584
## kscaled   0.739756    0.09626       7.685   2.157e-07    -97353705        1535
```

This is not as good a result as the best so far obtained, so we will not pursue it further.

20.7   Commentary

There are several ideas that are important to obtaining a decent solution with reasonable efficiency:

- It appears to be important to check if the differential equation integration was successful. An awkwardness in doing so is the fact that different methods in **deSolve** use different values of the flag to indicate success or lack thereof.

- Bounds may be useful in keeping parameters in a particular range, that is, positive, but may slow down the solution process.

The following are some points to consider:

- Is it worth figuring out how to generate analytic derivative information? This may be a quite difficult task.

- Should we bother with gradient methods when we do not have analytic gradients for problems such as this?

- Is our code more comprehensible if we explicitly include the exogenous tspan and data in the dot variables?

While we have found solutions here, it appears that drastically different sets of model parameters give rise to similar fits, as can be partly seen from the following table. This is a concern, as we then have to worry about the meaning of our parameters. That they are not well defined suggests that the model structure needs adjustment, rather than the solution method.

```
## Function value LLode * 1e-8 and parameters
## Guess,         223.44
##      Bi      Bw escaled kscaled
##  0.7500  0.7500  1.0000  1.1212
## ML - Nelder,  2.0806
##      Bi      Bw escaled kscaled
## 0.30733 1.14238 0.45378 0.96585
## Optimx-best,  2.0775
##            Bi      Bw escaled kscaled
## bobyqa 0.30601 1.8785 0.28343 0.98172
## nlfb-nlls,  2.1871
##      Bi      Bw escaled kscaled
## 0.33059 0.14624 1.26505 0.73976
## attr(, "pkgname")
## [1] "nlmrt"
```

Reference

Tien JH and Earn DJD 2010 Multiple transmission pathways and disease dynamics in a waterborne pathogen model. Bulletin of Mathematical Biology 72(6), 1506–1533.

21

Miscellaneous nonlinear estimation tools for R

This chapter concerns a number of R tools that are extensions or accessories to the materials we have discussed so far. Because they are treated here at the end of the book does not mean that they are unimportant. However, my concerns have been the machinery to find estimates of nonlinear parameters by optimizing functions. A number of the tools in this chapter stress other aspects of statistical estimation that illuminate the data or models in other ways.

21.1 Maximum likelihood

As maximum likelihood estimation is such a common task in computational statistics, several tools and packages exist for carrying out some of the forms of ML tasks.

mle in the **stats4** (part of the base R distribution, R Core Team (2013), but it appears that one needs to load it with require(stats4)) is intended for minimizing a function minuslogl using a method chosen from optim(). This tool appears to have fallen into disuse, possibly because it seems to be rather fragile. However, it does compute the solution for our Hobbs maximum likelihood example introduced in Chapter 12, and we include an illustration of use of fixed parameters (masks).

```
require(stats4, quietly = TRUE)
lhobbs.res <- function(xl, y) {
    # log scaled Hobbs weeds problem -- residual base parameters on log(x)
    x <- exp(xl)
    if (abs(12 * x[3]) > 50) {
        # check computability
```

```
        rbad <- rep(.Machine$double.xmax, length(x))
        return(rbad)
    }
    if (length(x) != 3)
        stop("hobbs.res - - parameter vector n!=3")
    t <- 1:length(y)
    res <- x[1]/(1 + x[2] * exp(-x[3] * t)) - y
}
lhobbs.lik <- function(Asym, b2, b3, lsig) {
    # likelihood function including sigma
    y <- c(5.308, 7.24, 9.638, 12.866, 17.069, 23.192, 31.443)
    y <- c(y, 38.558, 50.156, 62.948, 75.995, 91.972)
    xl <- c(Asym, b2, b3)
    logSigma <- lsig
    sigma2 = exp(2 * logSigma)
    res <- lhobbs.res(xl, y)
    nll <- 0.5 * (length(res) * log(2 * pi * sigma2) + sum(res * res)/sigma2)
}
mystart <- list(Asym = 1, b2 = 1, b3 = 1, lsig = 1) # must be a list
amlef <- mle(lhobbs.lik, start = mystart, fixed = list(lsig = log(0.4)))
amlef
```

```
##
## Call:
## mle(minuslogl = lhobbs.lik, start = mystart, fixed = list(lsig = log(0.4)))
##
## Coefficients:
##      Asym       b2       b3      lsig
##   5.27911  3.89366 -1.15976 -0.91629
```

```
amlef@min # minimal neg log likelihood
```

```
## [1] 8.117
```

```
amle <- mle(lhobbs.lik, start = as.list(coef(amlef)))
amle
```

```
##
## Call:
## mle(minuslogl = lhobbs.lik, start = as.list(coef(amlef)))
##
## Coefficients:
##      Asym       b2       b3      lsig
##   5.27914  3.89373 -1.15976 -0.76712
```

```
val <- do.call(lhobbs.lik, args = as.list(coef(amle)))
val
```

```
## [1] 7.8215
```

```
# Note: This does not work
print(lhobbs.lik(as.list(coef(amle))))
```

```
## Error: 'b2' is missing
```

```
# But this displays the minimum of the negative log likelihood
amle@min
```

```
## [1] 7.8215
```

The function mle2() from package **bbmle** (Bolker and Team, 2013) does seem to work more satisfactorily in my opinion. It has a data= argument that allows us to specify the data with which the model functions are to be computed. Moreover, the standard output includes the value of the likelihood.

```
require(bbmle, quietly = TRUE)
lhobbs2.lik <- function(Asym, b2, b3, lsig, y) {
    # likelihood function including sigma
    x1 <- c(Asym, b2, b3)
    logSigma <- lsig
    sigma2 = exp(2 * logSigma)
    res <- lhobbs.res(x1, y)
    nll <- 0.5 * (length(res) * log(2 * pi * sigma2) + sum(res * res)/sigma2)
}
y0 <- c(5.308, 7.24, 9.638, 12.866, 17.069, 23.192, 31.443)
y0 <- c(y0, 38.558, 50.156, 62.948, 75.995, 91.972)
mystart <- list(Asym = 1, b2 = 1, b3 = 1, lsig = 1) # must be a list
flist <- list(lsig = log(0.4))
amle2f <- mle2(lhobbs2.lik, start = mystart, data = list(y = y0),
fixed = flist)
amle2f


##
## Call:
## mle2(minuslogl = lhobbs2.lik, start = mystart, fixed = flist,
##     data = list(y = y0))
##
## Coefficients:
##      Asym       b2       b3      lsig
##   5.27911  3.89366 -1.15976 -0.91629
##
## Log-likelihood: -8.12

amle2 <- mle2(lhobbs2.lik, start = as.list(coef(amlef)), data = list(y = y0))
amle2


##
## Call:
## mle2(minuslogl = lhobbs2.lik, start = as.list(coef(amlef)),
## data = list(y = y0))
##
## Coefficients:
##      Asym       b2       b3      lsig
##   5.27914  3.89373 -1.15976 -0.76712
##
## Log-likelihood: -7.82
```

Note: The log likelihood displayed above for amle2 and amle2f is the negative of the function we have minimized. The function value is found as

```
amle@details$value

## [1] 7.8215
```

maxLik offers a somewhat different set of tools for maximum likelihood estimation. Its optimizers (some of which are built on the optim() function) are all **maximizers**, in contrast to almost all the methods in this book, which by default minimize functions. Unfortunately, one quirk of this package is that it does not respect the quietly=TRUE directive of the require() function, so I had to be a bit more aggressive in suppressing the messages that did not fit the formatting of this page. Note that the objective function not only is specified as the log likelihood (rather than its negative) but that the parameters are supplied to this function as a vector. Personally, I prefer to supply parameters as a vector, but other users may find the individual parameters aid in linking the computations to their particular problems.

```
suppressMessages(require(maxLik))
llh <- function(xaug, y) {
    Asym <- xaug[1]
    b2 <- xaug[2]
    b3 <- xaug[3]
    lsig <- xaug[4]
    val <- (-1) * lhobbs2.lik(Asym, b2, b3, lsig, y)
}
aml <- maxLik(llh, start = c(1, 1, 1, 1), y = y0)
aml

## Maximum Likelihood estimation
## Newton-Raphson maximisation, 15 iterations
## Return code 2: successive function values within tolerance limit
## Log-Likelihood: -7.8215 (4 free parameter(s))
## Estimate(s): 5.2791 3.8937 -1.1597 -0.76715
```

likelihood (Murphy, 2012) uses simulated annealing to maximize the likelihood function and is more at home in Chapter 15 where it has been mentioned. Moreover, this package uses a very different structure for maximum likelihood estimation than the other packages discussed in this section, so I will not pursue its use further here.

21.2   Generalized nonlinear models

There are always generalizations of any style of model or computation, and R developers have not been slow to pursue such possibilities.

Package **nlme** is a very large software collection for linear and nonlinear mixed effect models. We have already seen how various packages are available for nonlinear least squares in Chapter 6, but those tools assume we should minimize the sum of squared residuals. The residuals can be explicitly adjusted by weights, or these can be passed to the computation by the weights argument to the nls() function. However, it is assumed that the residuals are uncorrelated. When they have a variance/covariance structure, we need to modify the objective function. In the **nlme** function gnls(), this is precoded for us.

Package **gnm** (Turner and Firth, 2012) aims to allow the fitting of models that are rather different, and specified in a different way, from those we have presented in the rest of the book. In particular, **gnm** considers overparameterized representations, where we have more parameters and modeling terms than we actually need. Our goal is to determine which are useful and generate an effective model with a subset of the model features. (This description is my own; the package authors would probably express things otherwise.) There is both a manual and a quite extensive vignette for the package. We can run the Bates version 16.7 of the logistic model with **gnm** and put the `nlxb()` solution after for comparison.

```
require(gnm, quietly = TRUE)
y0 <- c(5.308, 7.24, 9.638, 12.866, 17.069, 23.192, 31.443)
y0 <- c(y0, 38.558, 50.156, 62.948, 75.995, 91.972)
t0 <- 1:12
Hdta <- data.frame(y = y0, x = t0)
formula <- y ~ -1 + Mult(1, Inv(Const(1) + Exp(Mult(1 + offset(-x), Inv(1)))))
st <- c(Asym = 200, xmid = 10, scal = 3)
ans <- gnm(formula, start = st, data = Hdta)

## Running main iterations.....
## Done

ans

##
## Call:
## gnm(formula = formula, data = Hdta, start = st)
##
## Coefficients:
## Mult(., Inv(Exp(Mult(1 + offset(-x), Inv(1))) + Const(1))).
##                                                       196.19
## Mult(1, Inv(Exp(Mult(. + offset(-x), Inv(1))) + Const(1))).
##                                                        12.42
## Mult(1, Inv(Exp(Mult(1 + offset(-x), Inv(.))) + Const(1))).
##                                                         3.19
##
## Deviance:            2.5873
## Pearson chi-squared: 2.5873
## Residual df:         9

require(nlmrt, quietly = TRUE)
anls <- nlxb(y ~ Asym/(1 + exp((xmid - x)/scal)), start = st, data = Hdta)
anls

## nlmrt class object: x
## residual sumsquares =  2.5873  on  12 observations
##      after  5     Jacobian and  6 function evaluations
##   name       coeff       SE      tstat      pval       gradient      JSingval
## Asym       196.186      11.31     17.35   3.167e-08   -3.726e-09      44.93
## xmid       12.4173      0.3346    37.11   3.715e-11   -1.492e-08      15.6
## scal       3.18908      0.0698    45.69   5.767e-12   -2.818e-08      0.0474
```

In this section, I have deliberately left out some packages that are distributed privately and do not satisfy the checks of CRAN. Unfortunately, while such software may contain useful resources, I do not want to suggest the use of tools that may introduce conflicts with the mainstream R packages.

21.3   Systems of equations

R has two packages that I consider well developed for estimating models that are specified by several equations: **sem** (Fox et al. 2013) and **systemfit** (Henningsen and Hamann, 2007).

systemfit (Henningsen and Hamann, 2007) is, to my view, more suited to multiequation models from econometrics. These are mostly specified as sets of linear models with exogenous and endogenous variables. While there is a form of nonlinearity introduced by the multiple equation nature of the models, the nonlinear optimization tools of this book are generally not used by practitioners in this field. Indeed, while the package includes a function called `nlsystemfit()`, the authors caution that it is still under development.

Most of the applications of **sem** are also likely to be collections of linear models. There is a useful vignette (Fox, 2006).

systemfit and **sem** use slightly different ways of specifying the model equations to their estimating functions. While both allow equations to be provided as R expressions, **sem** also allows, via a `specifyModel()` function, for a form of shorthand for the model. Different optimizers can be specified to fit the model, and I have found that I was able to create a modified **sem** and reinstall it with a modified optimizer. In a problem sent to me by John Fox to investigate a convergence issue, I changed one of the optimization tools from the `optim()` method CG (a code based on my own work but which I do not feel should now be used) to method BFGS.

21.4   Additional nonlinear least squares tools

While the central engines for solving nonlinear least squares tasks have been dealt with in Chapter 6, R has several tools to extend that functionality. We conclude with a brief mention of some of these.

nlstools (Baty and Delignette-Muller, 2013) contains a mixed bag of tools and data that reflect the agricultural research background of its authors. There are several functions for working with growth curves, as well as some aids for computing confidence intervals of parameters.

`fitdistr()` from package **MASS** (distributed with R) is designed to allow for fitting of common univariate distributions by maximum likelihood. Here the package developers have done the work for us of writing the objective function and in some cases of providing the starting values for optimization. Bounds can be specified on the parameters. **fitdistrplus** (Delignette-Muller et al. 2013) is a package that extends this by allowing for censored data and by permitting other criteria such as moment matching to be used for estimating the distribution parameters. (This, of course, is moving away from optimization.)

In a similar manner, the authors of package **grofit** (Kahm et al. 2010) have provided some precanned routines for fitting a variety of growth curves that are common in biological models. Like other packages of this type, the features provided are intended for a particular audience and not necessarily the easiest for others to employ. For example, we need to find the sum of squares buried in the object nls within the returned solution. Here is an example using the familiar Hobbs data. The summary() output for the returned object (called ah here) is, to my eyes, not very helpful. Therefore, I examined what the object contains and found the nls() component, which is displayed in two ways. Also given is the model component that tells us a logistic model was found to be "best" by the function gcFitModel(), which actually tries several forms and compares them using the Akaike information criterion (AIC). This is defined as

$$\text{AIC} = 2 * (k + \text{nll}) \tag{21.1}$$

where k is the number of parameters in the statistical model and *nll* is the minimized value of the negative log-likelihood function for the estimated model. Smaller is better.

```
## tgrofit.R -- Use Hobbs problem to test grofit
y <- c(5.308, 7.24, 9.638, 12.866, 17.069, 23.192, 31.443)
y <- c(y, 38.558, 50.156, 62.948, 75.995, 91.972)
tt <- 1:12
require(grofit, quietly = TRUE)
ah <- gcFitModel(time = tt, data = y)

## --> Try to fit model logistic

## ....... OK

## --> Try to fit model richards

## ....... ERROR in nls(). For further information see help(gcFitModel)

## --> Try to fit model gompertz

## ....... OK

## --> Try to fit model gompertz.exp

## ... OK


print(summary(ah))

##    mu.model lambda.model A.model integral.model stdmu.model stdlambda.model
## 1     15.38       6.0391  196.19          376.9      0.5832         0.20539
##    stdA.model ci90.mu.model.lo ci90.mu.model.up ci90.lambda.model.lo
## 1     11.307            14.42           16.339               5.7013
##    ci90.lambda.model.up ci90.A.model.lo ci90.A.model.up ci95.mu.model.lo
## 1                6.377          177.59          214.79           14.236
##    ci95.mu.model.up ci95.lambda.model.lo ci95.lambda.model.up ci95.A.model.lo
## 1            16.523               5.6366               6.4417          174.02
##    ci95.A.model.up
```

```
## 1          218.35

ah$model

## [1] "logistic"
```

summary(ah$nls)

```
##
## Formula: data ~ logistic(time, A, mu, lambda, addpar)
##
## Parameters:
##         Estimate Std. Error t value Pr(>|t|)
## A        196.186     11.307    17.4  3.2e-08 ***
## mu        15.380      0.583    26.4  7.8e-10 ***
## lambda     6.039      0.205    29.4  3.0e-10 ***
## ---
## Signif. codes:  0 '***' 0.001 '**' 0.01 '*' 0.05 '.' 0.1 ' ' 1
##
## Residual standard error: 0.536 on 9 degrees of freedom
##
## Number of iterations to convergence: 6
## Achieved convergence tolerance: 3.07e-07
```

ah$nls

```
## Nonlinear regression model
##    model: data ~ logistic(time, A, mu, lambda, addpar)
##     data: parent.frame()
##      A       mu lambda
## 196.19  15.38   6.04
##   residual sum-of-squares: 2.59
##
## Number of iterations to convergence: 6
## Achieved convergence tolerance: 3.07e-07
```

21.5   Nonnegative least squares

A rather different tool is provided in **nnls** (Mullen and van Stokkum, 2012). Here we wish to solve a least squares problem that looks to be like a usual linear least squares calculation. However, we require that the resulting parameters to be positive and possibly that there sum be scaled. Such problems arise in decoding spectra, in various imaging calculations (where the pixel cannot have a negative intensity), and some other domains.

To keep the presentation simple, we will fabricate a small example. Suppose we have three substances that have known spectra signals given as Sig1, Sig2, and Sig3 below. These form our dictionary matrix. We have a mixture of these, with proportions or concentrations 0.23, 0.4, and 0.1. We will not impose a sum constraint but simply assume the combined signal is given by the linear combination of the columns of the dictionary matrix. We can measure the signal (spectrum) of the combination, but there is some measurement error or "noise" which we add to simulate a real problem.

```
Sig1 <- c(0, 0, 1, 2, 3, 4, 3, 2, 1, 0, 0, 2, 4, 8, 16, 8, 5, 1, 0)
Sig2 <- c(2, 3, 4, 2, 1, 1, 3, 5, 7, 1, 0, 0, 0, 0, 0, 0, 5, 0, 0)
Sig3 <- c(0, 0, 0, 0, 0, 4, 4, 4, 1, 0, 0, 1, 14, 18, 16, 18, 15, 1, 10)

C <- cbind(Sig1, Sig2, Sig3)

bb <- C %*% as.matrix(c(0.23, 0.4, 0.1))

scale <- 0.15
require(setRNG)

## Loading required package: setRNG
setRNG(kind = "Wichmann-Hill", seed = c(979, 1479, 1542),
normal.kind = "Box-Muller")
d <- bb + scale * rnorm(19)
```

Now that we have our problem, let us solve it with **nnls**.

```
require(nnls, quietly = TRUE)
aCd <- nnls(C, d)
aCd

## Nonnegative least squares model
## x estimates: 0.23301 0.40724 0.10194
## residual sum-of-squares: 0.55
## reason terminated: The solution has been computed sucessfully.
```

However, we can use other methods, including nonlinear least squares and minimization methods that include bounds constraints. First, let us define our residual, Jacobian, sum-of-squares, and gradient functions. And then, as a reminder that it is always a good idea to do so, we check them, but here I have commented out the printout of results.

```
############# another example ############
resfn <- function(x, matvec, A) {
    x <- as.matrix(x)
    res <- (A %*% x) - matvec
}

jacfn <- function(x, matvec = NULL, A) {
    A
}

ssfun <- function(x, matvec, A) {
    rr <- resfn(x, matvec, A)
    val <- sum(rr^2)
}
ggfun <- function(x, matvec, A) {
    rr <- resfn(x, matvec, A)
    JJ <- jacfn(x, matvec, A)
    gg <- 2 * as.numeric(t(JJ) %*% rr)
}
```

We could check the functions by running the following code.

```
# Check functions:
xx <- rep(1, 3)
resfn
print(resfn(xx, d, C))
jacfn
print(jacfn(xx, d, C))
ssfun
print(ssfun(xx, d, C))
ggfun;print(ggfun(xx, d, C))
```

Using nlfb() from package **nlmrt** is straightforward.

```
require(nlmrt, quietly = TRUE)
strt <- c(p1 = 0, p2 = 0, p3 = 0)
aCdnlfb <- nlfb(strt, resfn, jacfn, lower = 0, matvec = d, A = C)
aCdnlfb

## nlmrt class object: x
## residual sumsquares =  0.55031  on  19 observations
##     after  3    Jacobian and  4 function evaluations
##    name      coeff        SE       tstat      pval     gradient  JSingval
## p1      0.233008    0.01645     14.16  1.803e-10  -3.98e-11     43.11
## p2      0.407237    0.01602     25.41  2.309e-14  -2.976e-12    11.57
## p3      0.101936    0.009332    10.92  7.932e-09   6.769e-11    10.07
```

Finally, let us try some optimization tools. We note that starting on the bounds does not yield a solution for bobyqa() and nmkb(). The latter code specifically says NOT to start on the bound, but this is not mentioned for bobyqa(), although this outcome is not totally surprising to me given that this code does use some heuristics that might not always work. We get solutions when we alter the start. Also note how we would call the optimizers without a gradient function.

```
require(optimx, quietly = TRUE)
strt <- c(p1 = 0, p2 = 0, p3 = 0)
strt2 <- c(p1 = 0.1, p2 = 0.1, p3 = 0.1)
lo <- c(0, 0, 0)
aop <- optimx(strt, ssfun, ggfun, method = "all", lower = lo,
  matvec = d, A = C)

## Loading required package: numDeriv
##
## Attaching package: 'numDeriv'
##
## The following object is masked from 'package:maxLik':
##
##    hessian
## Warning: no non-missing arguments to max; returning -Inf
## Warning: no non-missing arguments to min; returning Inf
## Warning: nmkb() cannot be started if any parameter on a bound

summary(aop, order = value)
```

```
##             p1      p2      p3      value fevals gevals niter convcode  kkt1
## Rcgmin  0.23301 0.40724 0.10194 5.5031e-01     12      4    NA        0  TRUE
## Rvmmin  0.23301 0.40724 0.10194 5.5031e-01    170     21    NA        0  TRUE
## L-BFGS-B 0.23301 0.40724 0.10194 5.5031e-01      9      9    NA        0  TRUE
## nlminb  0.23301 0.40724 0.10194 5.5031e-01     25     20    19        0  TRUE
## spg     0.23301 0.40724 0.10194 5.5031e-01     15     NA    13        0  TRUE
## hjkb    0.23302 0.40724 0.10193 5.5031e-01    338     NA    19        0 FALSE
## bobyqa       NA      NA      NA 8.9885e+307    NA     NA    NA     9999    NA
## nmkb         NA      NA      NA 8.9885e+307    NA     NA    NA     9999    NA
##            kkt2 xtimes
## Rcgmin     TRUE  0.004
## Rvmmin     TRUE  0.016
## L-BFGS-B   TRUE  0.000
## nlminb     TRUE  0.000
## spg        TRUE  0.160
## hjkb       TRUE  0.016
## bobyqa       NA  0.000
## nmkb         NA  0.000
```

```
aop2 <- optimx(strt2, ssfun, ggfun, method = "all", lower = lo, matvec = d, A = C)
summary(aop2, order = value)
```

```
##             p1      p2      p3   value fevals gevals niter convcode  kkt1
## Rcgmin  0.23301 0.40724 0.10194 0.55031     12      4    NA        0  TRUE
## Rvmmin  0.23301 0.40724 0.10194 0.55031     18      8    NA        0  TRUE
## nlminb  0.23301 0.40724 0.10194 0.55031     26     20    19        0  TRUE
## L-BFGS-B 0.23301 0.40724 0.10194 0.55031      9      9    NA        0  TRUE
## spg     0.23301 0.40724 0.10194 0.55031     18     NA    14        0  TRUE
## bobyqa  0.23301 0.40724 0.10194 0.55031     89     NA    NA        0  TRUE
## hjkb    0.23301 0.40724 0.10194 0.55031    373     NA    19        0 FALSE
## nmkb    0.23304 0.40729 0.10191 0.55031    109     NA    NA        0 FALSE
##            kkt2 xtimes
## Rcgmin     TRUE  0.004
## Rvmmin     TRUE  0.000
## nlminb     TRUE  0.000
## L-BFGS-B   TRUE  0.000
## spg        TRUE  0.164
## bobyqa     TRUE  0.004
## hjkb       TRUE  0.012
## nmkb       TRUE  0.016
```

```
## No gradient func-
tion is needed -- example not run aopn<-optimx(strt, ssfun,
## method='all', control=list(trace=0), lower=c(0,0,0), matvec=d, A=C)
## summary(aopn, order=value) aop2n<-optimx(strt1, ssfun,
method='all', lower=lo,
## matvec=d, A=C) summary(aop2n, order=value)
```

21.6   Noisy objective functions

In Section 1.4.4, we mentioned that some optimization problems have objective functions that can only be imprecisely computed. The obvious example is some process where the value of this "function" is actually measured, such as the time taken for vehicles to complete journeys under some settings of parameters believed to control that measurement. A more statistical example arises when we must integrate over a distribution function to compute a log likelihood or similar objective function. By using Monte-Carlo methods, we can do the integration quickly but imprecisely. Parameter values apart from the estimates are of limited interest, so the question then arises as to whether it is better to compute the integral precisely and optimize the resulting function, or to use an optimization method that is tolerant of imprecise objective function values. This was explored by Joe and Nash (2003), and we produced a Fortran code as well as a description of it.

The Joe–Nash program is of the "model and descend" type. We sample the function (we referred to it as a response surface) at a number of points, estimate a quadratic model or paraboloid, and move to the minimum of the model. Then we repeat this process. Of course, the details are very messy:

- The parameters are bounded to avoid wild excursions out of the region of interest, as the imprecision may suggest unfortunate trial sets of parameters.

- The sampling strategy, that is, location and number of points, can be very important, yet difficult to formulate.

- It is important to drop some points from the set used to build a model, and once again the rules for dropping points are important but are not simple to specify.

While this procedure should be relatively straightforward in R compared to Fortran, we have yet to build a package to do this. Note Deng and Ferris (2006) for an approach based on Powell's UOBYQA optimizer. As both **minqa** and **nloptr** packages have versions of Powell's codes, it might be possible to adopt this strategy for use in R, but I have not yet seen attempts to do this.

21.7   Moving forward

This concludes my treatment of nonlinear parameter optimization with R. The story is not, of course, ended. As I have been writing, there have been discussions with others about many developments in this field. Unfortunately, some are not sufficiently stable to be sensibly discussed yet.

In the next few years, I anticipate that there will be a serious debate about the optimization tools in R and how they are structured. This is, of course, a common issue for open source software projects, and R is one of the largest and most successful of these. For R, the debate will need to deal with the difficulty of shifting away from legacy tools that are no longer being developed and that have limitations that are awkward to work around easily. An example is the nls(), part of the base distribution. Its writers have largely dropped out of R activities, it has, as we have seen, a tendency to fail more than it needs to. Similarly, optim() is showing its age. As we have also seen, there are replacements that address some of the weaknesses, but readers should expect these to have their own cycle of adoption and then replacement with yet other packages.

The more positive message of such ongoing change is that needs do generate new packages. The process is a little messy, in that the new offerings may work wonderfully for some problems and be weak with others. Over time, however, the collective experience advances the system.

References

Baty F and Delignette-Muller ML 2013 nlstools: tools for nonlinear regression diagnostics. R package version 0.0-15.

Bolker B and Team RDC 2013 bbmle: Tools for general maximum likelihood estimation. R package version 1.0.13.

Delignette-Muller ML, Pouillot R, Denis JB, and Dutang C 2013 fitdistrplus: Help to fit of a parametric distribution to non-censored or censored data. R package version 1.0-1.

Deng G and Ferris MC 2006 Adaptation of the UOBYQA Algorithm for Noisy Functions. WSC '06: Proceedings of the 38th Winter Simulation Conference, pp. 312–319. Winter Simulation Conference, Monterey, CA.

Fox J 2006 TEACHER'S CORNER structural equation modeling with the sem package in R.

Fox J, Nie Z, and Byrnes J 2013 sem: Structural Equation Models. R package version 3.1-301.

Henningsen A and Hamann JD 2007 systemfit: a package for estimating systems of simultaneous equations in r. Journal of Statistical Software 23(4), 1–40.

Joe H and Nash JC 2003 Numerical optimization and surface estimation with imprecise function evaluations. Statistics and Computing 13(3), 277–286.

Kahm M, Hasenbrink G, Lichtenberg-Fraté H, Ludwig J, and Kschischo M 2010 grofit: fitting biological growth curves with R. Journal of Statistical Software 33(7), 1–21.

Mullen KM and van Stokkum IHM 2012 nnls: The Lawson-Hanson algorithm for non-negative least squares (NNLS). R package version 1.4.

Murphy L 2012 Likelihood: methods for maximum likelihood estimation. R package version 1.5.

R Core Team 2013 R: A Language and Environment for Statistical Computing. R Foundation for Statistical Computing Vienna, Austria.

Turner H and Firth D 2012 Generalized nonlinear models in R: an overview of the GNM package. R package version 1.0-6.

Appendix A

R packages used in examples

The following R packages were used in the examples in this book. Sources for those not on CRAN (Comprehensive R Archive Network) are noted.

- **adagio**: Discrete and global optimization routines
- **alabama**: Constrained nonlinear optimization
- **BB**: Solving and optimizing large-scale nonlinear systems
- **bbmle**: Tools for general maximum likelihood estimation
- **compiler**: This is now provided as a standard package in the base distribution.
- **DEoptim**: Global optimization by differential evolution
- **deSolve**: General solvers for initial value problems of ordinary differential equations (ODE), partial differential equations (PDE), differential algebraic equations (DAE), and delay differential equations (DDE)
- **expm**: Matrix exponential
- **GA**: Genetic algorithms
- **gaoptim**: Genetic algorithm optimization for real-based and permutation-based problems
- **GenSA**: R functions for generalized Simulated Annealing
- **gnm**: Generalized nonlinear models
- **grofit**: The package was developed to fit many growth curves obtained under different conditions
- **linprog**: Linear programming/optimization

Nonlinear Parameter Optimization Using R Tools, First Edition. John C. Nash.
© 2014 John Wiley & Sons, Ltd. Published 2014 by John Wiley & Sons, Ltd.
Companion Website: www.wiley.com/go/nonlinear_parameter

- **MASS**: Functions and data sets to support Venables and Ripley, 'Modern Applied Statistics with S' (4th edition, 2002). This recommended package generally is installed with the base distribution.

- **maxLik**: Maximum likelihood estimation

- **microbenchmark**: Submicrosecond accurate timing functions

- **minpack.lm**: R interface to the Levenberg–Marquardt nonlinear least squares algorithm found in MINPACK, plus support for bounds

- **NISTnls**: Nonlinear least squares examples from NIST

- **nleqslv**: Solve systems of nonlinear equations

- **nlmrt**: Functions for nonlinear least squares solutions

- **nloptr**: R interface to NLopt

- **nloptwrap**: Wrapper for package nloptr

- **nls2**: Nonlinear regression with brute force

- **nnls**: The Lawson–Hanson algorithm for nonnegative least squares (NNLS)

- **numDeriv**: Accurate numerical derivatives

- **optextras**: A set of tools to support optimization methods (function minimization with at most bounds and masks), https://r-forge.r-project.org/R/?group_id=395

- **optimx**: A replacement and extension of the `optim()` function

- **pander**: An R pandoc writer

- **polynom**: A collection of functions to implement a class for univariate polynomial manipulations

- **pracma**: Practical numerical math functions

- **profr**: An alternative display for profiling information

- **quantreg**: Quantile regression

- **rbenchmark**: Benchmarking routine for R

- **Rcgmin**: Conjugate gradient minimization of nonlinear functions with box constraints

- **RcppDE**: Global optimization by differential evolution in C++

- **reportr**: A general message and error reporting system

- **rgenoud**: R version of genetic optimization using derivatives

- **Rmalschains**: Continuous optimization using memetic algorithms with local search chains (MA-LS-Chains) in R

- **Rmpfr**: R MPFR – multiple precision floating-point reliable
- **rootoned**: Roots of one-dimensional functions in Ronly code https://r-forge.r-project.org/R/?group_id=395
- **Rsolnp**: General nonlinear optimization
- **Rvmmin**: Variable metric nonlinear function minimization with bounds constraints
- **setRNG**: Set (normal) random number generator and seed
- **smco**: A simple Monte-Carlo optimizer using adaptive coordinate sampling
- **soma**: General-purpose optimization with the self-organizing migrating algorithm
- **stats4**: Statistical functions using S4 classes. This collection of materials is part of the base distribution but appears to need loading via `require()`.
- **trust**: Trust region optimization.

Index

Nonlinear Parameter Optimization Using R Tools, First Edition. John C. Nash.
© 2014 John Wiley & Sons, Ltd. Published 2014 by John Wiley & Sons, Ltd.
Companion Website: www.wiley.com/go/nonlinear_parameter

Printed and bound by CPI Group (UK) Ltd, Croydon, CR0 4YY

02/10/2023

08123923-0001